FAO
FISHERIES
TECHNICAL
PAPER

338

An introduction to monitoring, control and surveillance systems for capture fisheries

prepared by
P. Flewwelling
Consultant

Food
and
Agriculture
Organization
of
the
United
Nations

Rome, 1994

M-43
ISBN 92-5-103584-9

PREPARATION OF THIS DOCUMENT

This paper is intended as a technical reference for all Fisheries Administrators. It is hoped that it will contribute towards efforts to implement coastal and offshore fisheries management schemes using appropriate strategies of monitoring, control and surveillance (MCS) wherever they are needed.

Mr Peter Flewwelling is President of Ocean Resources Limited, 29 Crimson Drive, Lower Sackville, Nova Scotia, Canada, B4C 3L2.

ACKNOWLEDGEMENTS

Special thanks are given to the staff of FAO who took time and effort to offer their advice and experience in the design and review of this paper. Further, thanks and sincere appreciation for the support and assistance from the official reviewers from the South Pacific Forum Fisheries Agency; the Organization of Eastern Caribbean States; the Seychelles Fishing Authority; Mr. Winston Miller, the Chair of the Belize Fisheries Advisory Board and former Fisheries Administrator; and to Peter Derham, a well known fisheries manager with extensive international experience. My thanks go also to the Canadian International Development Agency (CIDA), the International Centre for Ocean Development (ICOD) and Roche International for permission to use their graphics and materials from other projects. Finally, my sincere appreciation to all those unofficial advisors who took the time to review and offer advice during the development of this paper.

DISTRIBUTION

FAO Fisheries Department
FAO Legal Office
FAO Regional Fishery Officers
Directors of Fisheries
Selected Institutes and Projects
Regional and International Fisheries Organizations

Flewwelling, P.
An introduction to monitoring, control and surveillance for capture fisheries
<u>FAO Fisheries Technical Paper</u> No. 338. Rome, FAO. 1995. 217 p.

ABSTRACT

The paper has been designed as a handbook for Fisheries Administrators (a term used to denote those fisheries authorities responsible for decisions regarding the fisheries of their country) for their consideration when establishing, or enhancing, monitoring, control and surveillance (MCS) systems in support of fisheries management initiatives. The paper is divided into three main parts: definition and context of an MCS system; design considerations for MCS; and, MCS operational procedures. Annexes provide detailed examples of operational issues to be addressed in MCS system design and implementation.

TABLE OF CONTENTS

LIST OF ACRONYMS

ADB	ASIAN DEVELOPMENT BANK
ASEAN	ASSOCIATION OF SOUTHEAST ASIAN NATIONS
CARICOM	CARIBBEAN COMMUNITY
CECAF	FAO FISHERY COMMITTEE FOR THE EASTERN CENTRAL ATLANTIC
CFRAMP	CARICOM FISHERIES RESOURCE ASSESSMENT AND MANAGEMENT PROGRAM
CIDA	CANADIAN INTERNATIONAL DEVELOPMENT AGENCY
CONVENTION	1982 UNITED NATIONS CONVENTION ON THE LAW OF THE SEA
DFAS	DEPARTMENT OF AGRICULTURE AND FISHERIES OF SCOTLAND
EEC	EUROPEAN ECONOMIC COMMUNITY
EEZ	EXCLUSIVE ECONOMIC ZONE
ETA	ESTIMATED TIME OF ARRIVAL
ETD	ESTIMATED TIME OF DEPARTURE
FAD	FISH AGGREGATING DEVICE
FAO	FOOD AND AGRICULTURE ORGANIZATION OF THE UNITED NATIONS
FFA	SOUTH PACIFIC FORUM FISHERIES AGENCY
GPS	GLOBAL POSITIONING SYSTEM
ICOD	INTERNATIONAL CENTRE FOR OCEAN DEVELOPMENT
ILO	INTERNATIONAL LABOUR ORGANIZATION
IMO	INTERNATIONAL MARITIME ORGANIZATION OF THE UNITED NATIONS
IOFC	INDIAN OCEAN FISHERIES COMMISSION
IOMAC	INDIAN OCEAN MARINE AFFAIRS CO-OPERATION PROGRAM
IPTP	INDO-PACIFIC TUNA PROGRAM
IRCS CALL	INTERNATIONAL TELECOMMUNICATIONS UNION RADIO SIGNS
ITQ	INDIVIDUAL TRANSFERABLE QUOTA
ITU	INTERNATIONAL TELECOMMUNICATIONS UNION
LOA	LETTER OF AGREEMENT
LUX-DEV.	DUCHY OF LUXEMBOURG DEVELOPMENT PROGRAM
MARPOL 73/78	INTERNATIONAL CONVENTION ON THE PREVENTION OF POLLUTION FROM SHIPS
MCS	MONITORING, CONTROL AND SURVEILLANCE
NAFO	NORTHWEST ATLANTIC FISHERIES ORGANIZATION
OECS	ORGANIZATION OF EASTERN CARIBBEAN STATES
SFA	SEYCHELLES FISHING AUTHORITY
SOLAS	INTERNATIONAL CONVENTION ON SAFETY OF LIFE AT SEA
SQ3	FLAG SIGNAL - "STOP OR HEAVE TO, I AM GOING TO BOARD YOU"
UK	UNITED KINGDOM

UNCED	UNITED NATIONS CONFERENCE ON ENVIRONMENT AND DEVELOPMENT
UNDP	UNITED NATIONS DEVELOPMENT PROGRAMME
USA	UNITED STATES OF AMERICA
USSR	FORMER UNION OF SOVIET SOCIALIST REPUBLICS
VMS	VESSEL MONITORING SYSTEM

1. INTRODUCTION

The Food and Agriculture Organization of the United Nations has been assisting fisheries administrations in developing countries for several years to address the issues they confront in managing and developing fisheries. Assistance often takes the form of specialist advice on policy formulation and on strategies for management and development. This paper is an extension of that assistance.

Growing concerns of many developing countries include their current inability to control fisheries in their Economic Zones, the apparent high cost to do so, and the failure of some of the developed country fisheries which were formerly used as examples of fisheries management. The latter failures appear to have resulted from difficulties in implementing appropriate fisheries management strategies. In many countries there are no established systems for monitoring, control and surveillance of maritime zones to conserve their marine fisheries and associated habitats. Such countries need assistance in addressing this situation, to reap the potential benefits for their citizens from appropriate fisheries and marine resource management.

Although there is much written about MCS, there is no single document addressed to the Fisheries Administrator, (a term used to denote the fisheries authorities responsible for decisions regarding the fisheries in their country) with guidance and advice on how to introduce or strengthen the MCS capability of their country, sub-region or region. This technical paper is intended to fill this void and, consequently, is addressed to senior fisheries managers, primarily in developing countries. It focuses on MCS for coastal and offshore fisheries with attention to both national and foreign-owned fleets. This paper discusses: (i) the definition and context of MCS; (ii) design considerations for the components of an MCS system; and (iii) MCS operational procedures. Annexes provide additional details and examples of operational issues.

2. OVERVIEW

Monitoring, control and surveillance (MCS) have often been seen as self-explanatory and not requiring much consideration, as it is perceived as "simply the policing of the various maritime zones controlled by the State" - is it not? This technical document attempts to clarify this erroneous view of MCS and demonstrate how monitoring, control and surveillance are important to the implementation of any oceans related policy, in particular for fisheries management.

There are many influencing factors and global initiatives which have brought the subject of monitoring, control and surveillance to the fore in the international and national fora. The coming into force of the 1982 United Nation Convention on the Law of the Sea in November of 1994 has recognized the benefits of the Convention and once again raised the profile of the obligations States have with respect to the assessment of their fishing stocks, the allocation of the surplus to national needs to third parties and the further obligation to conserve their fisheries, including the fisheries habitat. The following excerpts from articles of the Convention are reminders of these obligations:

Article 61

Conservation of the living resources

1.   The coastal State shall determine the allowable catch of the living resources in its exclusive economic zone.

2.   The coastal State, taking into account the best scientific evidence available to it, shall ensure through proper conservation and management measures that the maintenance of the living resources in the exclusive economic zone is not endangered by over-exploitation. As appropriate, the coastal State and competent international organizations, whether subregional, regional or global, shall co-operate to this end.

3.   Such measures shall also be designed to maintain or restore populations of harvested species at levels which can produce the maximum sustainable yield, as qualified by relevant environmental and economic factors, including the economic needs of coastal fishing communities and the special requirements of developing States, and taking into account fishing patterns, the interdependence of stocks and any generally recommended international minimum standards, whether subregional, regional or global.

Article 62

Utilization of the living resources

1.   The coastal State shall promote the objective of optimum utilization of the living resources in the exclusive economic zone without prejudice to article 61.

2.   The coastal State shall determine its capacity to harvest the living resources of the exclusive economic zone. Where the coastal State does not have the capacity to harvest the entire allowable catch, it shall, through agreements and other arrangements and pursuant to the terms, conditions, laws and regulations referred to in paragraph 4, give other States access to the surplus of the allowable catch, having particular regard to the provisions of articles 69 and 70, especially in relation to the developing States mentioned therein.

Article 73

Enforcement of laws and regulations on the coastal State

1.   The coastal State may, in the exercise of its sovereign rights to explore, exploit, conserve and manage the living resources in the exclusive economic zone, take such measures, including boarding, inspection, arrest and judicial proceedings, as may be necessary to ensure compliance with the laws and regulations adopted by it in conformity with this Convention.

Article 192

States have the obligation to protect and preserve the marine environment.

Article 194

Measures to prevent, reduce and control pollution of the marine environment

1.   States shall take, individually or jointly as appropriate, all measures consistent with this Convention that are necessary to prevent, reduce and control pollution of the marine environment from any source, using for this purpose the best practical means at their disposal and in accordance with their capabilities, and they shall endeavour to harmonize their policies in this connection.

Article 197

Co-operation on a global or regional basis

States shall co-operate on a global basis and, as appropriate, on a regional basis, directly or through competent international organizations, in formulating and elaborating international rules, standards and recommended practices and procedures consistent with this Convention, for the protection and preservation of the marine environment, taking into account characteristic regional features.

The Convention details enforcement responsibilities for each of these obligations.

Further, the United Nations Conference on Environment and Development (UNCED) in Brazil in 1992 and the resultant Chapter 17 of Agenda 21 emphasized the urgent requirement to preserve our marine/fisheries environment on a global basis. The initiatives of several agencies on controlling pollution at sea (MARPOL 73/78); the United Nations Environment Programme's Global Environmental Facility Fund and Regional Seas Project and port State control initiatives for the merchant fleet, are all contributing towards global standards to protect our marine resources. The FAO flagging initiative to assist in the control of fishing vessels on the high seas, the potential of extending port State control and international co-operation with respect to fishing vessels, the efforts at responding to the need to establish standards for safety-at-sea for fishing vessels, and the most recent efforts at establishing standards for the control of fishing on the high seas of highly migratory species and straddling stocks, are all symbolic of the growing global trend and concern regarding fisheries resources and their habitat. Finally, there is also an initiative to establish a "code of conduct", one might call it, for responsible fishing practices. This is all applicable to fisheries and the fisheries habitat.

As these Conventions come into force and the initiatives gain credibility, so too do the obligations of participating States to implement these agreements to which they are signatories. Many States have, and are, reaping the benefits of these agreements as internationally respected legal instruments, but now as they come into force, a case in point being the Convention, there is a realization that the implementation of these agreements carry considerable obligations for each State. Monitoring, control and surveillance (MCS), especially for the fisheries sector, is the implementing tool to meet these obligations. MCS, which has often been thought of as a luxury for developed States, has now become an obligation for all States to work together to conserve the marine resources and their environment.

As an initial step, many countries are now focusing on developing and establishing policies and strategies to meet their obligations under the 1982 United Nations Convention on the Law of the Sea.

The ideal situation, and a logical approach to implement the terms of the Convention, would be to develop an oceans policy which would establish government priorities and the strategy for the conservation and sustainable use of all marine resources within the zone. From this oceans policy would flow the integrated oceans planning and management framework under which fisheries management would be developed. Most countries do not have the luxury of this longer term development process and have chosen to track in parallel these two initiatives; oceans policy development on the one hand, and fisheries management, including MCS strategies, on the other. Consequently, although countries remain cognizant of the fact that fisheries management must link to overall oceans policy when it is ultimately established, the development of the MCS systems required to implement fisheries management plans must go forward to address the more immediate need to conserve fish stocks and their habitat. The parallel tracking of the development of oceans policy and fisheries MCS initiatives will impact on the latter as they are implemented, most probably by forcing the multiple tasking of fisheries MCS resources to address other related, maritime priorities. This impact will also depend on the focus of political will and the priority of fisheries in the overall oceans strategy. These points will obviously serve as an incentive, in the first instance, for countries to develop the most cost effective and appropriate MCS system to meet fisheries requirements, with the flexibility to assume other, secondary, fisheries-related tasking, such as pollution and environmental monitoring, as delegated by the government.

Fisheries management and the resultant MCS activities will depend on the decisions of government as to the level of its involvement in the fishing industry. Government can assume a very intrusive role, such that it actually runs the industry, controls potential income of fishers, and micro-regulates the harvesting sector. Alternatively, it can maintain a less intrusive role, whereby the fishing industry accepts its responsibilities and role within the framework of general government conservation principles and legislation. The South Pacific Forum Fisheries Agency has adopted this latter style of fisheries management. In either case, there are several perceptions regarding the fishers and MCS activities which have negatively coloured fisheries management, and fisheries MCS itself. There is the idea that many of the problems with fisheries in each country stem from the illegal fishing activities of foreign fishing fleets. Although foreign fishing fleets have had, and continue to have, an impact on fisheries conservation efforts in the majority of cases, the greater impact on fisheries in each country usually stems from its own domestic fishing industry in the coastal and nearshore fishing zones. Fisheries MCS activities focused solely on the control of foreign fishing efforts will therefore undoubtedly be less than successful as a key component in the conservation of the resources, if the domestic fishing effort is not also being addressed.

Another erroneous perception with respect to MCS is that it is a simple exercise of "activating the military might of the State to arrest the foreign poachers". Activating the military infrastructure for MCS is an initiative many countries have taken in the past and found to be prohibitively expensive and politically sensitive. Further, MCS activities are not solely enforcement operations, but also include the collection of data, and quality control of that data, for input into the stock assessment, social and economic, and enforcement exercises that comprise the components of fisheries management as well as safety-at-sea. This is not the normal use of military enforcement machinery and is neither applicable, nor appropriate, for domestic fishers.

The expense of MCS activities is usually the primary concern of any government in designing and implementing a system of controls. Cost effectiveness and efficiency are key to successful MCS operations. The surveillance, or enforcement, aspect of fisheries MCS has proven to be most cost effective and responds best to the political sensitivity of international fisheries incidents when it remains a civil action, without the potential involvement of the military might of the nation. Further, on the matter of expense, military machinery costs much more to build and operate than equivalent civilian equipment for a civil action. As an example, this can be verified by those who have attempted to place a civilian radio system on a piece of military equipment. There are significant additional costs associated with such an initiative in order to make the equipment meet the appropriate standards for military acceptance. Another cost consideration is whether a country requires a large armour-plated, heavily crewed vessel with associated high firepower and costs to carry out fisheries patrols and exercise civil police duties of data gathering and enforcement. The latter function can be exercised for all fisheries in a fully satisfactory manner with significant cost savings using a civilian vessel with fewer crew, less armour and firepower and hence, lower fuel consumption and general costs.

The basic needs for fisheries MCS normally include vessels that can remain at sea with the fishing fleets, appropriately equipped aircraft for cost-effective rapid surveillance and information collection and, an adequate coastal support infrastructure for verification of landings and the monitoring of the port trade of fish products.

Effectiveness of operations can be enhanced considerably if only one ministry is responsible, or has the lead role and authority, for the implementation of MCS activities, instead of two or three. This significantly reduces the lines of communications for the command and control, to use military terminology, of the monitoring and surveillance components for MCS activities, making them more efficient and responsive in a timely manner. Further, bureaucracy is usually significantly reduced using vessels under one ministerial control, as opposed to a combined civilian and military operation. These points are not to suggest that a country should completely obviate the use of military equipment as an option where no other options are available, but the military option is one which has fairly considerable consequences in terms of cost and control.

It has also been found through the experience of both developed and developing countries that MCS operations can be expedited with reduced costs and in an effective manner through co-operation with neighbouring countries on bilateral, sub-regional or regional initiatives. Examples of this include initiatives of the South Pacific Forum Fisheries Agency (FFA) and the Organization of Eastern Caribbean States (OECS) and initiatives being established in West Africa. Considerable cost savings can also be realized through the appropriate use of licenses as a control tool, as a source of basic information for management planning, and as an alternative to a free access fishing scheme while minimizing the cost to the resource owners, the taxpayers. As a practical example, licenses can require all transhipment of fish to occur in port, where they can be monitored much more accurately and safely, thereby contributing to MCS in a cost-effective manner. Fishing vessels could be encouraged to use ports by a reduction in administrative and bureaucratic requirements compared to regular transport and shipping vessels. This could also enhance the opportunity of input into port State control initiatives which States may wish to initiate in the future.

These types of cost saving strategies contribute to the implementation of "no [minimal] force" MCS strategies to the benefit of the coastal State.

A point made for years by fisheries officers is the issue of creating unenforceable legislation. Unenforceable legislation, or that which is not understood nor acceptable to the fishers, rapidly destroys the credibility and support for a government in its efforts to conserve its fisheries resources. Such legislation usually results in active subversion of its intended benefits by the fishers and the fishing industry. One example is the prohibition of catching a certain species of fish. Unless the individual is seen in the act, it is almost impossible to enforce. The prohibition would be best stated as it being against the law to be found in possession of this certain species of fish above a certain acceptable bycatch or tolerated level.

A somewhat controversial example of fisheries management systems which are difficult to enforce in developing countries is the legislation creating individual transferable quotas (ITQs). Although this system has all the positive attributes of reducing periodic fishing pressures on fish stocks, it is very difficult to enforce without sophisticated communications and data networks backed up by an expensive, real time, accurate and verifiable monitoring and surveillance system. This management scheme may be appropriate for developed countries, but could prove very difficult to implement in developing countries. Experience has shown that the ITQ management scheme may be appropriate and manageable in small island states with both small populations and fishing fleets, and where the fishery is largely export oriented. To date the system has not been found appropriate in other situations, especially where the fishery has a domestic component.

An initiative which will result in a near, real time vessel monitoring system is being developed by the FFA for use in the South Pacific. Until this system is perfected and acceptable to the fishers, it is suggested that such a quota management system may be found to be administratively difficult to operate and result in a continuation of overfishing.

It has been found over time that no MCS activities will be successful if there is not an understanding and acceptance by fishers of the rationale behind the MCS actions being implemented. Legislation which is unenforceable denigrates the credibility of the entire fisheries management process in the eyes of the supposed benefactors, the fishers, and consequently quickly establishes a "them and us" confrontational attitude on both sides. It must be remembered that most fishers are realists who are fishing for their survival and "whilst sometimes libellously assumed by the ill-informed to be crooks, are perhaps best described as being as honest as the next man, but hard, individualistic businessmen running very competitive and often highly capitalized operations. It is worth remembering that they do so in the face of a largely unforgiving sea that creates a working environment which, in terms of industrial health and safety, [has one of the worst industrial accident rates in the world]: and they operate increasingly in an economic climate of ever increasing overheads countered only by the proceeds of catches which [are] subjected to [greater] quota restriction. All of this is done in the knowledge that the success of their venture and the livelihood of their crews depends entirely on their individual skill, effort and initiative. Given these pressures, it is perhaps not surprising that such independent minds do not always take kindly to bureaucratic controls, especially if these appear to them to have little practical purpose".[1]

3.   MONITORING, CONTROL AND SURVEILLANCE: DEFINITION AND CONTEXT

3.1   Definition of MCS

Along with the potential for enhanced economic benefits, the 1982 Convention on the Law of the Sea also brought with it responsibilities for coastal States in the utilization of resources in EEZs. It is this latter responsibility that, in many cases, has demonstrated the need for development and control over the use of a country's marine resources. Fisheries are central to this development, as fish and their habitat are key resources in the exclusive economic zone. Although the objectives of fisheries management and MCS are generally to take advantage of the economic opportunities of the extension of the EEZ, they also include the exercise of sovereign rights over the zone, conservation of marine resources, and collection of appropriate data on activities to ensure sound, rational oceans and fisheries management planning. Fisheries MCS needs to be defined in light of these points.

There is ample literature on the subject of MCS and there are several definitions and interpretations; those commonly used by fisheries personnel stem from the MCS Conference of Experts in 1981 in Rome and are broadly defined as:

(i) **monitoring** - the continuous requirement for the measurement of fishing effort characteristics and resource yields.

(ii) **control** - the regulatory conditions under which the exploitation of the resource may be conducted.

(iii) **surveillance** - the degree and types of observations required to maintain compliance with the regulatory controls imposed on fishing activities.[2]

More simply stated, MCS is the implementation of a plan or strategy. In the case of oceans management and fisheries, it includes the implementation of operations necessary to effect an agreed policy and plan for oceans and fisheries management.[3] MCS is an often

---

[1]   Derham, Peter (1987)

[2]   *FAO (1981)*

[3] *There is reference in this paper to oceans policy, integrated oceans planning and management, and fisheries management. This is to emphasize the fact that any management policy or plan which applies to oceans will impact on several components of integrated ocean management. Fisheries departments, as the central component in ocean resource management, have an obligation to assume a key role in conservation and protection of ocean resources. MCS then becomes multi-sectoral, being essential for implementation of fisheries management strategies, as well as necessary for the successful implementation of wider oceans policies, plans and strategies.*

overlooked aspect of oceans and fisheries management; but, in reality, it is key to the success of any planning strategy. The absence of a strategy and methodology for implementation of monitoring, control and surveillance operations would render a fisheries management scheme incomplete.

3.2   MCS components

There are three main components to MCS which, depending on cost, commitment, and organizational structure (national, sub-regional, or regional), will be configured uniquely for each system. These are the land, sea and air components. The latter now often includes the use of satellite technology.

The *land* component, or base of operations, can serve the inland, freshwater, and coastal aspects of fisheries monitoring, control and surveillance. The land component is usually the coordinating sector of all MCS activities and regulates the deployment of available resources to best address the changing situations in the fisheries. The coastal/land component is the sector responsible for port inspections and the monitoring of transhipments and trade in fish products to ensure compliance with fisheries legislation.

The *sea* component of MCS includes the actual technology for surveillance of the national, sub-regional or regional maritime zones of control. This component can include radar and vessel platforms which are utilized for these purposes. Traditional apprehension of an alleged violator of the laws which apply in an EEZ requires a " laying of hands" on the offender, mainly for the legal formality of arresting, but also for identification and securing of evidence. As this is sometimes costly in terms of vessels, crews and supplies, many nations are now favouring "no force" surveillance techniques. These include the use of observers, national or regional vessel registers or licenses and agreements which include clauses regarding the responsibility of the flag State for the actions of its citizens and vessels. The fisheries management strategy, if it utilizes zonal fishing divisions, quotas necessitating catch monitoring, mesh and gear restrictions and minimum/maximum fish sizes, requires vessels to carry fisheries personnel which can remain at sea with the fishing fleets to perform these functions, as they cannot be expedited from land. The patrol vessels also serve a maritime safety function while at sea.

The *air* component of MCS is usually the first level of response to a coastal state/regional concern in its area of responsibility or interest. The flexibility, speed and deterrence of air surveillance makes it a very useful and cost-effective tool for fisheries management. This component also provides the cheapest and most rapid information collection on fishing effort in the zone of interest, from either the aircraft or satellite platforms. The cost of these systems is directly related to the sophistication of the technology utilized. Air surveillance, while providing initial information regarding the activity in the fisheries, can also be the first indicator of potential illegal activity in the zone. This latter information is the base on which further MCS action can be precipitated. Air surveillance also has the added advantage of a secondary tasking capability for fisheries habitat and general coastal zone management monitoring.

The potential benefits to seafarers in difficulty at sea, pollution, habitat and general coastal zone management monitoring are significant in that they can also be addressed during fisheries MCS activities. Multi-tasking of expensive fisheries MCS resources for other fisheries-related monitoring functions can be cost efficient and also effective in terms of integrated ocean management programmes, in particular with respect to fisheries and the marine environment.

Acceptance of the aforementioned definition of monitoring, control and surveillance emphasizes the point that MCS is a three-tiered system. A common error in establishing MCS systems throughout the world is the concentration on the "S" in MCS, or the enforcement phase, without due regard to the importance of the other two phases. The monitoring and control aspects of MCS provide the base of information and legal framework for sound fisheries management and operational planning. The surveillance phase is the most expensive aspect of MCS and hence, developing countries must look at the most cost-effective methods to carry out functions related to this component.

It is useful to remember that the most beneficial aspect of enforcement is preventive enforcement, or voluntary compliance. It is similar to car repairs where it is less expensive to conduct preventive maintenance than to carry out time consuming, and costly, repairs. The same principles apply to MCS. An understanding and acceptance by those who will most benefit from MCS actions, the fishers, can prove most effective in gaining their support, commitment and involvement for surveillance purposes. Voluntary compliance has the added advantage of then permitting a focus for the expensive enforcement resources towards areas of major national, sub-regional or regional concern in the most cost-effective manner. The deterrent impact contained in the control phase, or the legislation supporting MCS, will determine the potential repetition of offenses. If there is an understanding of the rationale for the fisheries legislation, this will disuade the fishers from violating these acts and regulations. This idea can be reinforced through the drafting of enforceable legislation and appropriate penalties.

3.3   Role in fisheries management

Where does MCS fit in fisheries management? The schematic of fisheries management on the next page helps address this question.[4] In essence, this schematic assumes that there are three linked components of fisheries management:

1. collection of data on biological, economic, social aspects of the fisheries and basic information on the fishers, boats and gear;

2. decision making or fisheries management planning; and,

3. implementation, the MCS aspect of fisheries management involving both government officials and members of the fisheries community and industry.

The first component (data gathering) includes collection and analysis of biological/resource assessment data, basic fisheries information on fishers, boats and gear,

---

[4] *Fanning, Paul J., Fisheries Management System, CARICOM Fisheries Newsnet, Volume 2, Number 2, CARICOM Fisheries Resource Assessment and Management Program, Belize, C.A., August 1993.*

fishing trends and patterns, and social and economic data for the harvesting, processing and marketing sectors of fisheries. The analysis of these data provides the input into the fisheries planning exercise.

The second component (decision-making), includes the entire process of consultation and negotiation with all parties which influence the decisions concerning a fishery. This should result in fisheries management plans for the harvesting, processing and marketing sectors. The key factor in this decision-making component which affects the entire process of fisheries management is the political will and its commitment to sustainable and rational fisheries management. Political commitment is crucial to success in the implementation of fisheries management plans.

The last component, which is often the most difficult for governments to deal with due to potentially high cost or other government priorities or arrangements in the fisheries sector, is the implementation of these plans. It is represented by monitoring, control and surveillance of the fisheries and the fishers, and is an absolute requisite to the successful implementation of fisheries management plans. The most comprehensive and acceptable plan on paper will not result in successful fisheries management unless it is implemented through the use of monitoring, control and surveillance operations. Lack of attention or commitment to implementation of MCS activities often results in overfishing, collapse of resources and economic loss to future generations.

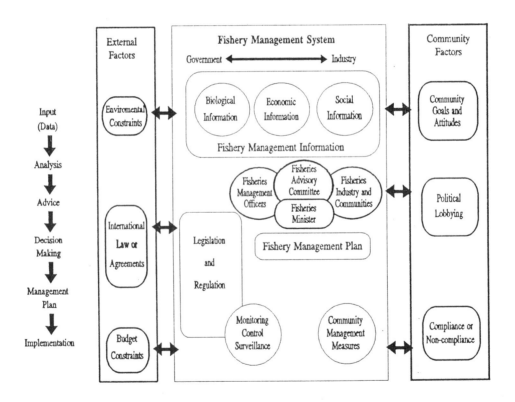

In the past, officials often reflected the view that fisheries management includes only the biological studies for resource assessment and development of management plans, and there the process ends. The fish were then expected to conserve themselves and the industry was to be the steward for processing and marketing. The support for data collection and policing of fishers and the industry to ensure appropriate input into future fisheries management plans, and the successful realization of these plans, was low. Now, there is a growing awareness of the declining condition of the environment and a greater acceptance of the need for investment in implementation (MCS) of natural resource management plans, including those for fisheries. It is in this vein that countries are attempting to address the earlier concerns with respect to the funding of MCS activities.

Experience in these initiatives to date has demonstrated a need for one agency to assume, or be assigned, the lead for MCS activities to prevent the confusion, duplication and associated inefficiencies and extra costs of multi-agency authorities. As fisheries have the greatest risk with respect to mismanagement of renewable marine resources and their habitat, it may be a consideration that fisheries departments be delegated this lead role in MCS matters. As noted earlier, however, MCS activities in support of fisheries management can also accommodate other secondary tasks which pertain to conservation of a nation's natural resources and its environment. The expensive infrastructure required for MCS activities, especially in terms of aircraft and vessels, when coupled with the overall responsibility for MCS of the coastal zone and EEZ of each country, or group of countries, for fisheries and their habitat/marine environment is easier to rationalize as being appropriate to the magnitude of the task.

4.   DESIGN CONSIDERATIONS FOR MCS

4.1   Influential factors

There are three groups of factors which may influence decisions respecting the type of MCS system required to meet the needs for fisheries management. These include the geographic, demographic, industry and international profile of the fishery, economic factors, and the political will and commitment related to analysis of the first two groups of factors.

4.1.1 Geographic, demographic, marine industry and international factors

Size of the EEZ and the Fishing Area within the Zone: The area of fishing of both the domestic and foreign fleets will have a significant influence on the design of the MCS system for each country. There will be a significant difference in the situation, and hence the MCS strategy, in the Philippines, where responsibility and authority for the artisanal fisheries (over 400,000 small fishing vessels called "bancas", with a zone out to 15 km around its 7,000 islands) has been delegated to the municipalities, compared to that for a country such as Equatorial Guinea, with 8,750 artisanal fishers and foreign fishers taking advantage of the State's inability to fund and operate appropriate MCS mechanisms.

The area of active fisheries may also raise control concerns, due to the migration of fish stocks between fishing areas and countries. This can also cause an artisanal versus offshore conflict, if there are incursions of the latter into the small boat fisher's area. This can become critical when the EEZ of a country has only a small rich fishing ground and the fishing pressures from all fishing sectors is intense.

Another consideration in this regard is the size of the various fisheries. If the domestic fishery is small, inshore stocks are strong, and the fishery is not a threat to conservation, then MCS activities can focus on data collection and reasonable control mechanisms required to maintain the health of the fishery. If the offshore fishery is extensive and lucrative, involving either domestic or foreign-owned vessels, then MCS activities to collect data, control the fishery, and patrol the area may require greater effort, to ensure that benefits of the resources are conserved for the State. On the other hand, if the artisanal fishery is overfished and there is growing pressure on the offshore fisheries, with little knowledge of the resource base, then consideration will need to be directed to greater information gathering on the latter and options for re-directing effort from the artisanal to the offshore fisheries in a cost-effective, controlled manner, if this is possible. A key factor in this scenario may be the targeted species, both offshore and inshore, offshore usually being high value species for profit while the artisanal fisheries focus on any fish, often lower value, for survival.

The strategy to establish an MCS system for a foreign fishing zone that is restricted outside 12 or 15 nautical miles is much easier to develop than one in which foreign fishermen are permitted in inshore zones when fishing for certain species. The latter necessitates a verification of fishing catches while the former only requires geographic confirmation. This is one example of the kind of strategic analysis which is required when developing an MCS plan appropriate for the fishery and the enforcement capability of the State.

Further, taking into consideration the above factors regarding the size of the fishing area, it is obvious that a country with a 200 kilometre (km) coastline and a 50 km wide continental shelf does not need the same MCS system as a country with a 1000 km coast and a 200 nautical mile wide shelf. The physical size of the EEZ and the active fisheries area within the zone will have a significant influence on the design of the MCS system. A large zone and fishing area may require air surveillance or other such infrastructure to patrol the areas of concern, whereas a narrow fishing zone might be surveyed cost effectively using other technology, possibly land-based, such as over-the-horizon radar, coast watch systems, or vessel monitoring systems (VMS), coupled with less expensive "no force" strategies. An example of this difference is the comparison of the 209,000 square kilometres of EEZ for Ghana, compared to the 20 million square kilometres of EEZ to be surveyed by the FFA for its member countries. The options for MCS strategies must be considerably different if the systems are to be cost effective and this depends largely on the size and geographical location of the fishing areas within these EEZs, or collective areas of responsibility.

The profile of the fishing fleets, domestic, foreign, artisanal and offshore, is a consideration for the implementation of MCS strategies. The age, condition, size, fishing capacity, gear type and fishing patterns of the vessels will all have an impact on what the State may wish the vessels to do to comply with its MCS policies. These factors may also pre-empt requirements to carry certain equipment, due to the state of the vessel, crewing, current equipment or other such factors which can increase costs to the point that fishing may not be viable, especially for the domestic fleet. On the other hand, there may be a requirement, after noting the profile of the fleets, to establish minimum safety and equipment standards not only for the well being of the fishers, but also to minimize the risk of pollution at sea.

Topography of the coastline: A coastline with several bays, river outlets and important mangrove habitat would require a more complicated MCS scheme to conserve fisheries resources than one with steep rocky cliffs and minimally important habitat. The difficulties for surveillance would also be more complex in the former case, due to the indentations in the coastline which would necessitate a physical presence to survey fishing activity, rather than radar or other, less expensive technology.

Other interests: The importance of tourism, the enhancement of industrial capacity, the requirements for ports and shipping, maritime safety-at-sea and pollution monitoring and controls can all have an impact on the strategy developed for fisheries management, with consequent repercussions on the MCS system adopted. There will be a requirement for discussion and liaison with appropriate ministries to ensure that government priorities are met and, in so far as possible, fisheries requirements and benefits to society are recognized and respected. Fisheries management priorities will often seem to conflict with priorities from tourism, industry, and marine transportation initiatives. It must be realized however, that the sustainable benefits of each of these industries, and the economic and employment situation for the State, will benefit from appropriate attention to, and conservation of, marine resources. There will be a need for a mechanism for discussion and resolution of differences in approaches and priorities for each of these important industries.

International pressures: International pressures from distant water fishing nations (DWFNs) and the short term economic benefits of foreign currency cash flow can be attractive to states that do not have fully developed fisheries, but often with the consequence of non-sustainable exploitation of their fisheries resources. DWFNs have their own difficult situations of over-capacity and recession in their economies and fishing industries which they have been trying to address. These countries may, in the past, have been be less appreciative of, or sensitive to, the negative impact of some of their agreements with the developing nations. On the other hand, it would have been unreasonable to expect that these fleets would negotiate an agreement which would have had them operating at a financial loss. Consequently, these agreements have in the past, unfortunately, been to the detriment of the conservation goals of coastal states, which did not, and still do not, have the economic flexibility to ensure the implementation of appropriate MCS systems to conserve their fisheries resources. Often, the only protection against uncontrolled overfishing and lack of compliance with regulatory measures due to shortages in resources and training, is through strong co-operation on a bilateral, sub-regional and regional basis with regard to MCS activities. Cooperative efforts can result in economic and international pressures against non-compliance with internationally respected, conservation principles which would not otherwise be achievable on a single state basis. The potential for the establishment of international standards and guidelines is greater in this forum of international and regional cooperation. Examples include the aforementioned initiatives such as; the flagging agreement, Code of Conduct for Responsible Fishing, port state control and the control of fishing for highly migratory species and straddling stocks.

In recognition of the rights and interests of all States under the Convention on the Law of the Sea, developing countries must address their obligations to deal with conservation concerns and, at the same time, international fishing partners must ensure compliance with national fisheries legislation and international conventions dealing with fisheries in the name of conservation.

One particular note of special pertinence today, is the economic temptation to register DWFN vessels in national registers when there is no capability to control the activities of the "new flag" vessels. Some of these "new flag" vessels operate with short term immediate interests in internationally sensitive areas of the world and without appropriate attention ot conservation. This brings international pressure on the flag State with respect to its credibility and commitment to internationally accepted fisheries conservation and principles and to the implementation of the Convention. Registration of these vessels should be avoided.

Involvement of fishers, communities, organizations, cooperatives, unions and fishing companies: It should be self-evident that the cooperation of the fishing industry and fishers is essential to cost effective fisheries management. If the industry, fishers and their communities and organizations actively participate, and are recognized as real participants, in fisheries management, MCS planning and related activities, there is a much greater potential for successful implementation of these plans, to the benefit of all. On the contrary, if the opposite transpires, it will be extremely expensive and very difficult to successfully implement any fisheries management plans. Lack of recognition, input, involvement and understanding of the principles and rationale behind the proclaimed fisheries management scheme has often resulted in non-compliance, alienation of the fisheries department officials and active subversion of the intended plan, thereby placing much more pressure on the need for surveillance. Most fishers wish to conserve the resources and will support such initiatives if the efforts are reasonable, enforceable and understandable.

MCS operations directed towards education and seeking input from fishers are facilitated if there are strong fishers' or community organizations in place to discuss these issues. The independence of fishers is well known, consequently there is often a reluctance for fishers to join together for these types of discussions. Assistance to fishers in getting them to recognize and accept the advantages of having a collective voice is one of the challenges of fisheries educational initiatives in seeking input and support for MCS activities.

Demographics of the domestic fishery: Other issues may impact the type of MCS strategy to be adopted to ensure the continued health of fisheries, especially the domestic fishery. One such example is in the Seychelles, where the demography of the fishery is such that the average age of fishers is very high, as other employment opportunities appear more appealing and lucrative to the younger population. This creates several unique challenges for the State to re-kindle an interest in fishing as a profession and to build a controlled fishery using Seychellois, instead of other international fishers. This also presents an opportunity to educate the new fishers in the benefits of fisheries conservation. Alternatively, if young Seychellois are not encouraged into the fishery, the MCS system design will need to focus on the possibility of an increase in foreign offshore fisheries.

4.1.2   Economic factors

Contribution of fisheries to the GNP

It is obvious that the contribution of the fishery to the national economy will determine its profile and the importance placed on fisheries management activities. It makes economic sense, however, that the benefits from the resources exceed the cost of their conservation. As seen in the Seychelles, the potential contribution to the national economy by the displacement of international fisheries with a corresponding increase in domestic fishers, or other offshore fishing arrangements, could enhance the current situation. The GNP factor will likely have a significant effect on both the scale and the design of an MCS strategy.

Foreign currency earnings

A factor of considerable importance to several developing states is the earning of foreign currency by permitting international access to the fishery. It is very unfortunate that certain distant water fishing fleets have exploited this need in a strict business sense, to meet their own financial and employment requirements, without due consideration for conservation, or reasonable returns, for the coastal States. In the case of a developing country, where trained resources and infrastructure are not available to ensure the implementation of adequate MCS strategies, this can become a serious problem in the conservation of their fisheries resources. This is sometimes further complicated with the offer of external financial incentives to subvert any MCS efforts that the State may wish to impose. Governments, both developing and developed, must ensure that the fishing opportunities granted for both domestic and foreign fleets result in appropriate levels of compensation, and that these funds benefit the State.

Employment opportunities

The employment opportunities which can result from enhancing the fishing potential of the coastal State are also a factor in the consideration of the MCS strategy. In a situation where there are seen to be advantages in the long term of displacement of international fishing fleets, this may require training of coastal or island State nationals who will eventually assume these fishing rights. Training of nationals could therefore be a component of the access agreement with third party fishing fleets.

The demonstration of an increase in benefits to the individual fishers, either as increases in income, or employment opportunities, will have a positive impact on the national economy, the strength of the fishing sector and support for government fisheries policies.

Linked with the above strategy of training of nationals could be the opportunity to ensure the implementation of appropriate safety-at-sea equipment and practices in accordance with the coming into force of the Protocol to the Torremolinos Convention 1977. This Protocol will bring fishing vessels under port State control with respect to safety certificates and early attention to this point can result in economies for future training. This initiative can also link to the development of a Code of Conduct for Responsible Fishing resulting in a new attitude and blend of fishers with conservation and safety high on their list of priorities.

Benefits to other ocean users

Recognizing that MCS should conserve fisheries resources *and their habitats*, there may be mutual benefits for other ocean users as well as the fishers, if appropriate liaison between these users and the fisheries department and relevant MCS strategies can be developed. For example, careful assessment and control of tourism development, assurance of non-destructive fishing practices, development of marine parks, use of mooring buoys to reduce the damage to coral reefs from ad hoc anchoring, etc. can all benefit both fishers and other ocean users. MCS strategies can be developed to include consideration for such activities.

Small island States are realizing the negative impact of excessive use of pesticides and the marine pollution resulting from uncontrolled industrial development. It is being noted that all land based activities on small islands eventually influence the marine environment and they have the potential to kill the very marine resources and habitat, including the coral reefs, which bring the tourists and foreign currency to the State.

On a positive note, Belize, in Central America, has been attempting to regulate tourism, the development of marine parks, and the fisheries, and has established appropriate surveillance initiatives to ensure the implementation of the required management plans.

Other secondary benefits from MCS activities outside of fisheries activities can also be factors in establishing the type and structure of an MCS system. MCS activities can also assist in addressing safety-at-sea for national and international seafarers and marine pollution, environmental matters.

Low cost protein

A further economic factor to be considered in the design of an MCS system is the requirement of the State for protein for its citizens. The MCS design can include the requirement for a percentage discharge of fisheries products in the coastal State for distribution, processing or further export. This could result in the direct provision of protein for the people, or enhance industry development in the fish processing sector and have resultant positive impacts on export earnings for the procurement of this protein.

Regional mechanisms of potential support

The ultimate influencing factor on design will be **cost.** Can a State rationalize the cost of the MCS system desired? Economic logic suggests that no state should expend greater funds on conservation than the potential economic benefits that can be gained by such activities. It is often the cost factor that influences the country to seek sub-regional and regional initiatives to meet MCS requirements in the oceans sector. This is significantly viable, especially where fish stocks are shared, language barriers are not an influencing factor, maritime jurisdictions are adjacent, boundary delimitations have been resolved and political thrusts are synonymous. The cost savings of cooperation in the implementation of MCS operations has been demonstrated in the successful regional fisheries programmes in place today in the South Pacific and the Caribbean Basin. The FAO "flagging agreement" and "port State control" initiatives are mechanisms which also draw on the regional and international co-operative approach and consequently, merit consideration by developing States.

Regional cooperation is not without its own problems, however, especially where the aforementioned factors are not complimentary. Many lessons can be learned from other international fisheries organizations such as the Northwest Atlantic Fisheries Organization (NAFO), the European Union, the Indo-Pacific Fisheries Commission (IPFC) and the Indian Ocean Fisheries Commission (IOFC).

4.1.3   Political will and commitment

The key behind any ocean policy, planning and management system, including that for fisheries, is the degree of political will and commitment to implement such a system. The following includes some of the factors which may form the base for decisions on the structure of the MCS system to be developed.

The economic profile, or potential thereof, of the fishery in the national economy will undoubtedly determine the level of support the MCS initiatives will receive from the government. A potentially lucrative domestic fishery, and the MCS activities required to protect it, will probably receive significant government attention. Nevertheless, it will be necessary to balance the long term benefits with the short and medium term benefits to maintain the political support which is key to the successful development and implementation of MCS systems. Some of these could include the establishment of a database for resource management, revenue from resource users through license fees and greater control of the resources from licensing and surveillance, which conserve fish stocks and hence, increase incomes of the fishers. It must be emphasized that the political will and commitment of the country is the key factor in the successful design and implementation of MCS systems for fisheries.

This set of influential factors and their relative importance to the political objectives of a country make fisheries management and the resultant MCS strategy unique to each country. Key observations from reviews of successful national and regional MCS include:

1. There are no [universally acceptable] models and each system [in operation] is in fact, adapted to the cultural, geographic, political and legal framework of the country or region.

2. The operational character of the system will depend on management decisions made.

3. The legal and policy considerations are always taken into account when establishing an MCS system.

4. The decision making power is always in the hands of the civilians, even on surveillance matters.

5. The national and regional MCS are complementary to each other.[5]

After assessing the geographic, demographic, economic and political aspects of the fisheries, one should reflect on those aspects which relate directly to the design and implementation of an appropriate MCS system. These include questions regarding the most cost-effective and efficient system for the State, the legal framework required and acceptable

---

[5] *Bonucci, Nicola (1992) GCP/INT/NOR Field Report 92/22 (En)*

to the fishers, coordination of ministries, training, infrastructure, organizational support mechanisms, and funding sources. The following section highlights these points, with examples of options which have proved effective in the past.

4.2   Organizational considerations

4.2.1 National, sub-regional and regional structures

A growing trend over the past several years has been the formation of groups of countries with common political and economic interests. An enhanced awareness of the environment and the requirement for the conservation of natural resources has brought recent activity and interest in the formation of associations and international agreements along regional lines. Again lessons can be learned from the existing international fisheries organizations noted earlier in this document including the South Pacific Forum, NAFO, OECS, the European Union and others.

The rich resource base in many developing countries, their less than favourable economic situation and inability to protect these resources and the desire of the international fishing industry to gain access to these resources, all emphasize the need for cooperation amongst developing fishing nations. There are proven cost savings which can be experienced through cooperation with respect to acquisition of MCS resources, training, and negotiating from a larger power bloc for reasonable compensation in return for access to underutilized resources. The international exchange of appropriate fisheries data for MCS and fisheries management purposes, harmonized legislation, extradition and port State agreements are all benefits which should encourage international affiliations within regional developing countries.

The decision with respect to establishing an MCS system on other than a national basis does, however, depend on several factors. These include whether there is an existing organization which will serve the purpose, whether there is the international political will of states in the area, common interests in fisheries which would benefit from such a liaison, common language and cultural ties, and whether differences can be overcome. The security of what may be considered as sensitive data and the potential to resolve internal concerns to present a common face to the outside world are all additional considerations which impact on this decision regarding international fisheries and MCS cooperation. Finally, the difference in economic situations of possible member countries and the cost sharing arrangements for support to an international organization are also critical factors to be addressed. Each country must balance their own advantages and disadvantages prior to making a commitment to regional cooperation.

It should be noted that, despite difficulties experienced in the past with such regional cooperation, there are examples which have demonstrated that concerns can be addressed and the resultant efforts have proven advantageous to member states of such organizations. Probably the most advanced of such organizations in the developing world is the South Pacific Forum Fisheries Agency, with sixteen members, in its second decade of cooperative fisheries management. Another example is the Organization of Eastern Caribbean States Fisheries Unit (involving eight countries), a sub-regional organization in the Caribbean Basin. A sub-regional initiative is commencing in the Indian Ocean through the Seychelles which will link with the regional tuna research programme in Sri Lanka. Efforts of the CARICOM Fisheries Resource Assessment and Management Programme (twelve countries in the Caribbean) is another example with considerable potential for success under CARICOM management. Efforts in East Asia with fisheries management and MCS are well advanced and recently, operational manuals have been developed. These organizations can provide additional information for the consideration future MCS strategies.

4.2.2 Roles and responsibilities

A decision with respect to international cooperation will not abrogate the State from its responsibility to establish appropriate structures internally to address fisheries MCS issues. As an MCS system is developed, it is expected that there will be interest from ministries other than fisheries for access and input into the priorities and tasking of the resources. The ministries responsible for environment, national defense and coast guard, customs and immigration are a few which can be expected. It has been noted that MCS surveillance resources are expensive and that multitasking could be cost effective and efficient. Experience has noted however, that too many priorities can result in the acquisition of capital equipment which does not meet any function appropriately, consequently, it is suggested that for fisheries MCS activities, coordination be with other ministries with fisheries-related interests, such as coastal zone management and the marine environment. There is also a very real requirement to recognize that the ministry, or department, with a considerable stake and interest in conservation and sustainable use of ocean resources and their habitat, is fisheries. There is also a need in any operation to ensure that there is one lead agency with the appropriate authority to make decisions. Although there are several ministries with interest in MCS and, consequently, there may be a need for a coordinating committee, there still needs to be one final authority for decisions on the deployment and priorities for MCS operations. Split operational "command and control", to use military phraseology, have not met with success in the past in military or civilian operations. It is recommended that the fisheries department be provided with the lead role and ultimate authority for ocean sector MCS activities, in consultation with other interested departments and ministries.

If this is not acceptable to governments, an alternative could be an alternating chair for the coordinating committee, but this is a less preferred option. It is always preferable for fisheries managers to have to report to only one superior to maximize efficiency in operational MCS activities. This role and authority, whichever strategy is selected, should be formalized in legislation to establish clarity in the event of prosecutions.

4.2.3 Core infrastructure requirements

It is impossible to make concrete suggestions which would quantify the MCS requirements of each situation, as they will be different for each system. It is possible, however, to make suggestions on core requirements and to leave the quantification to each Fisheries Administrator. For example, the *monitoring* component of MCS would be best served through the receipt of information from the licensing unit, sea going units for sea sightings and inspections, port inspections and air sightings for vessel identification, activity and location. These seemingly simple tasks will require a data network and communications system. The system can also include data on the fishers licenses, fishing gear, types of vessels, fishing patterns, fishers and community profiles with respect to dependency and earnings from fishing and any other fisheries management information required. These data can be used for verification of licensing conditions, catch and effort for resource assessment and sustainable fisheries management planning for the future. The accumulation of data will require a storage and analysis capability which, although it can be manual, is best achieved with computers. It will be necessary to determine the number of entry points for data and to establish a network capability so the resultant analyses can be redistributed to all fisheries offices. This means offices in major fisheries landing points, collection schemes for sea, land and air data, and a central office for analysis, distribution and operational decision-making.

The *control* component of MCS will necessitate the determination of appropriate and enforceable legislation required to implement the fisheries plans for the various fisheries. It will address the authorities of fisheries personnel, legal fishing activities, minimum terms and conditions for fishing, and penalties for non-compliance. The minimum conditions which a State may wish to implement could include vessel identification, catch and reporting requirements, conditions for transhipment, standard catch and effort log sheets, terms and living conditions for observers, local agents for international fishing partners, and flag State responsibility for their vessels.[6] The control component will link with the State's justice department and also necessitate the appropriate training of all personnel involved in enforcing the legislation, *including sessions where the assistance and advice of the judiciary is requested*. The infrastructure requirement for this component of fisheries MCS is a team of knowledgeable fisheries lawyers for both drafting of enforceable and appropriate legislation, and also for the legal follow-up which may be required to implement these laws. This component will also establish the various mechanisms, strategies and policies for the implementation of operations (MCS activities) to implement the fisheries management plans.

The *surveillance* component of MCS will require fisheries personnel who not only collect data for the monitoring aspect of MCS during their surveillance duties, but also have the appropriate equipment, operating funds and training to enforce the legislative mechanisms of fisheries management. These personnel will require direction and infrastructure from which to operate, be it land, sea or air facilities. This is the enforcement component of fisheries MCS and as such is usually the largest and most expensive activity to fund. It must be remembered that for international MCS activities, there is a requirement under the Convention on the Law of the Sea that all surveillance equipment be clearly marked and identifiable as on government service.

Article 111

Right of hot pursuit

5.   The right of hot pursuit may be exercised only by warships or military aircraft, or other ships or aircraft clearly marked and identifiable as being on government service and authorized to that effect.

---

[6] *FFA Report 93/55 (1993)*

This requirement is often addressed through large, highly visible, government markings on all surveillance equipment. This is often supplemented on sea-going vessels by a fisheries flag which also denotes government surveillance on fisheries duties. The design and use of the flag should be clearly stated in the fisheries legislation. All fisheries vessels, regardless of size, should fly this flag while on fisheries duties.

Equipment requirements include the following, noting that it could be more cost effective if they were shared with other enforcement agencies involved in fisheries and oceans-related MCS operations:

National headquarters

A central headquarters near the departmental decision makers for the coordination of fisheries operations. This headquarters, as well as having the offices for the administration of fisheries, would, ideally, be situated adjacent the operations room.

A **central operations room** where the current status of the fishing operations can be shown through maps, plots and computer enhancement is recommended. This centre would need offices and personnel, with communications to appropriate field offices and other enforcement agencies, and direct communications to the minister responsible for fisheries. This becomes the situation briefing and de-briefing room when a sensitive fisheries matter arises. The Fisheries Administrator should thus have the capability, through the equipment and information accessed from this centre, to show the situation to decision makers and obtain direction for timely responsive action. These centres can be staffed by as few as two persons trained in communications, computer access and display techniques.

The **communications system** would ideally have telephone and appropriate radio communications to all fisheries centres and mobile platforms in the field for both safety and control of operations. Some MCS systems also use satellite communications in their networks, but it is very expensive. The modern HF radio systems on the market today could possibly assist in keepng costs to a minimum without losing effectiveness.

The **computer data system** for licensing and vessel registration, if it is decided to use such a system for this basic information, is now affordable. There are several licensing and vessel registration systems in use today and it will be a decision of the Fisheries Administrator as to the system which will best meet the State's needs.[7]

It is anticipated that the procurement of other major MCS equipment will be coordinated from the central headquarters. The **air surveillance** requirements for MCS may appear expensive, but are still seen as the cheapest method to receive rapid surveillance information with respect to fishing and fish habitat information in the zone. As a minimum, the following equipment is recommended. It is highly desirable to have a twin engine turbo

---

[7] *CFRAMP in the Caribbean, spent considerable effort and cost in designing a very flexible computer licensing and vessel registration system for PC computers to meet the varying needs of the twelve participating, developing countries while preserving the core data for each system. CARICOM, in Guyana, has the intellectual property rights for this system.*

prop aircraft for over-sea flights for safety, endurance and low maintenance costs. As a minimum, this aircraft should have radio communication to base and directly to sea-going fisheries patrol vessels, with common marine frequencies. The navigation system of the aircraft will need to be accurate, for it will form the base for prosecution of area infractions if they proceed to court. It would be desirable to have an endurance capability of 4-6 hours at economical speed. The speed for transit should be reasonable to maximize the time in the assigned patrol area, but the aircraft should be able to go slowly enough at low levels to identify and photograph fishing vessels. Photographic equipment for the recording of vessel activities is necessary.

More expensive air surveillance platforms are available. Additional equipment could include navigational equipment which can merge with the photographic evidence for court purposes. Night lights and instruments for IFR (Instrument Flight Rating) flights are very desirable for surveillance purposes. Onboard computers linked to accurate navigation systems, communications systems, radar and photographic systems, and the capability to accommodate vessel monitoring systems, would result in a very technologically advanced air surveillance platform. This would however, be an expensive operational tool which might be inappropriate for budgets of developing countries. The expense could possibly be easier to accommodate through regional cooperation and shared use of the air surveillance equipment.

The choice and equipping of the aircraft will be dependent upon the cost of the airframe and the ongoing costs of operation and maintenance. These latter two factors are often lost in the considerations for air surveillance, but they are the most significant costs for the MCS air activities. Aircraft can cost from a few hundred to several thousands of dollars per air hour depending on the configuration of the craft and equipment. If at all possible, it is highly desirable to ensure that there is local access to appropriate training and equipment to permit long term maintenance of the aircraft.

The **sea-going requirements** will vary considerably between countries, depending on the MCS strategy. The primary consideration when considering the acquisition or use of patrol vessels should be cost-effectiveness and affordability for the primary task, fisheries surveillance. One golden rule for cost effectiveness is that the fisheries patrol vessels should have at least the same sea keeping capability as the fleet which they are monitoring. There may be a temptation to procure fast, expensive vessels; however, it must be remembered that the purpose of these vessels is to transport the authorized fisheries officials to the fishing vessel for inspection. Although desirable for a quick transit to the patrol zone, this capability must be balanced against the high fuel and maintenance costs for such machinery. There may be a requirement to be able to overtake a departing vessel where there are no other diplomatic arrangements in place, but this capability should not overshadow the need for staying at sea and cost effectiveness on a daily basis. Fast patrol vessels have a tendency, no matter how well trained the operator, to operate at near top speed making them expensive in fuel and maintenance, with long down times due to equipment wear and repairs. Economic considerations may also make vessel charters a viable option instead of purchases. In this manner maintenance becomes the concern of the contracting firm and not the government.

Coastal and nearshore vessels do not need to stay at sea for prolonged periods and, consequently, smaller patrol vessels with one or two days sea keeping capability, or rapid response shore-based craft, might serve the purpose in this latter case. These vessels would be best equipped with radar and communications systems. The latter should include both marine radios and an additional one to communicate with the air surveillance platforms. Equipment for boarding and an appropriate boarding craft are recommended. Most countries have found the fibreglass, V-hulled, rubber-sided, speed boat (Lucas and Avon types being popular) to be most effective for boarding. The boarding boat should have two outboard engines, or one inboard/outboard and a small outboard engine, for safety. The boarding boat requires communication equipment to remain in contact with the patrol vessel at all times.

The offshore fishery will require the largest, and hence the most expensive, sea-going platforms in the infrastructure for fisheries MCS. These vessels can range from deep hulled trawler type vessels to offshore oil supply vessels with helicopter landing facilities. The key in the choice is, again, the capital cost for the vessel and equipment and, equally important, the operating and maintenance costs. Large vessels, by their very nature, require considerable fuel and provisions to operate for extended periods at sea. It has been recommended that wherever possible the management strategy attempt to keep the need for these expensive sea-going platforms to a minimum, but it must be realized that they are necessary for most traditional fisheries management schemes. Most offshore vessels for fisheries would best be equipped with twin diesels of a dependable model, with trained engineers, up-to-date navigation equipment, radar, photography equipment and radio communications. The latter should be at least as per the inshore patrol vessel, preferably with back-up systems and ideally, linkages to the air surveillance platforms. These vessels are intended as boarding platforms and their regular duties should not require them to be heavily armed assault vessels. As noted earlier, their first role is as boarding platforms for the fisheries officer.

Field offices

This category includes area and regional offices within the country. Similar principles apply for larger international regional organizations.

Office space is required for the field staff and their supporting administration. The office should be equipped with communications equipment to maintain contact with the headquarters and also to maintain communication with staff while on patrol. A radio communication network is usually sufficient for these activities. The office also requires the capability to collect and transmit data to other offices for compilation and analysis and also, to receive results for the planning of operations. Ideally, this capability can be achieved through a computer system with communication to these other offices. Transportation is required for staff for patrol purposes, either along the coast, at sea or by land, and also along the rivers and lakes where there are active fishing operations. This transportation can range from small boarding type craft, to motorcycles, to other types of vehicles. It is highly recommended that staff patrol in pairs for safety and personal security.

It is assumed that the Fisheries Administrator will ensure that each field office has in its reference library certain documents for assistance in their duties. These include:

- current fisheries legislation, acts, regulations, notices and the gazette,
- departmental guidelines for MCS activities including those for prosecutions,
- copies of any applicable treaties or agreements between countries in the region,
- a set of charts with updated baselines, territorial seas, EEZ and any specifically noted areas for fisheries management,
- past fisheries cases, details and penalties for reference during the preparation of a case,
- safety procedures and guidelines for MCS.[8]

Each officer should have in their possession, at all times, pictured documentation which clearly identifies the individual as a government authorized fisheries officer. This requirement should also be in the fisheries legislation. Each officer requires communications equipment to maintain contact with their base of operations. Each officer must have the appropriate accoutrements to record findings during the patrol; e.g., a patrol book with clear identification of the owner and sequentially numbered pages. This notebook could be used in court proceedings for identification of events and as an aide memoir for the officer. It is essential that it is properly maintained. This latter point is addressed in training for officers, noted in Annex D.

A final item for careful consideration is the provision of firearms to staff. There are several considerations with respect to this matter but in general **firearms, if they can be avoided, are not recommended for fisheries MCS.** However, it is recognized that there are situations when it would be considered very dangerous for fisheries officials to conduct their business without adequate personal protection. These include the proclivity of firearms use in the country, in the fishing industry, and levels of illegal activity throughout the industry and at sea. It is also important to assess compliance trends in the fishing industry and the history of difficulties with fishers, both domestic and foreign, regarding the protection of fisheries staff. Where fisheries have become incontrolled and unmanageable, it has been found that fishers resort to other, less desireable and violent activites, thus making the protection of fisheries MCS staff a priority requirement. Where possible however, other means of protection, such as guard dogs or batons for land based fisheries personnel, are urged. The issuance, carriage and use of firearms should be considered as a tool for staff protection only; it is not recommended that firearms be considered or used as an aggressive surveillance tactic.

If it is therefore decided that firearms will be issued to staff, new considerations apply. The first of these is the appropriateness of the designated individuals for the carriage of the weapons. Not all persons are mentally suitable to carry and use firearms. The Canadian experience in fisheries recognized this fact and now a battery of psychological tests, as part of the selection process, are used to screen applicants for their suitability to carry firearms. Those found not suitable are released from further recruitment testing. The danger

---

[8] *Coventry, R.J., paraphrased from the South Pacific Forum Fisheries Agency Fisheries Prosecution Manual.*

in putting weapons in the wrong hands in terms of potential accidents can be significant and result in legal liability of the Department. The second major consideration is the initial training and the need for ongoing refresher training (at least annually). This is critical for the confidence of the staff in the proper use of the weapon and their own personal safety.

The decision on arming vessels for fisheries enforcement purposes is one which should not be taken lightly. It is a conscious decision to arm the vessel both for protection, *and potential aggressive action*. This decision may be necessary where fishing vessels commonly do not comply with the orders to halt for fisheries inspections. This scenario may result if no other enforcement strategies or agreements to ensure compliance have been established with the flag State of the vessel and diplomatic relations to address the situation are not available or have failed. In this case it may be the government's policy to permit aggressive, civil police action to apprehend the alleged offender. The matter of consequence here is the use of aggressive force. Legally, this force should be limited to that necessary to ensure compliance with the legal authorities, in this case the fisheries officer, or for protection of staff as appropriate. The amount of force that is deemed appropriate is always one of subjectivity, but excessive force may result in a case, if it goes to court, not being supported. It is suggested, therefore, that in the case of action involving the use of firepower, there be strict standards of application for the escalation of the use of force, from verbal warnings, through warning shots, to the use of force to physically stop the vessel; i.e., stop or potentially, sink the vessel. This action also includes the assumption that the boarding vessel is appropriately identified in accordance with the obligations under the Convention and that verbal identification has also been passed to the vessel being boarded. Countries should ensure that appropriate higher authorities are involved in the decisions to escalate the use of force. The use of force for protection when fired upon always remains with the master of the vessel to protect the ship and crew. The use of firearms by a fisher against a fisheries official or vessel, noting the identification above, should be responded in kind, with the maximum use of force directed at the source of firing.

It must be remembered that MCS surveillance is essentially a civilian police action, and not an aggressive military exercise hence, the use of military forces for fisheries MCS is recommended only for extreme cases.

This section has been presented to encourage reflection on the initial decisions regarding equipment to implement the MCS operational plan. The sum total is a requirement for a central control facility, a data system and network to the coastal areas and offices, offices along the coast for data collection and surveillance, communications and data collection facilities, appropriate transport equipment, land, sea and air platforms, and other safety equipment for surveillance operations. The requirement for appropriately trained directors, supervisors, and field personnel is assumed to be a central component of MCS.

4.2.4 Staffing

The requirements for MCS include personnel to address each of the components of monitoring, control and surveillance. The numbers of these personnel will vary with the MCS scheme in place, but basic requirements for the qualifications of these persons should remain fairly constant. These personnel need varied levels of expertise; for example, data collectors need to be literate, have good interpersonal skills and knowledge of the fishery and its policies and procedures. Sea-going experience is crucial to fisheries MCS, especially for sea operations. These individuals are often at the technician level. Observers for offshore vessels also fall into this category of staff. It must be noted here that the observer scheme is only appropriate if capable, honest and dedicated personnel are available, preferably with offshore sea experience. In many countries, close supervision is also necessary for this scheme to be productive due to the onboard pressures placed on these individuals. If these factors cannot be met, consideration for the implementation of an observer programme is inappropriate. Further points on this initiative are presented in the annex on observers.

The ability to analyze the data collected for fisheries management decisions and operational deployments requires a higher level of knowledge and competence, both academic and practical.

The control component requires individuals with comprehensive and very good knowledge of fisheries and the law. These individuals would work with the lawyer assigned from the Ministry of Justice for the design of enforceable laws and also for the internal decisions regardiing MCS operations.

Finally, there is the need for surveillance personnel. This includes personnel to operate small and large patrol vessels. Further, aircraft personnel will be required, as will the every day fisheries officials at the junior and more senior levels for coastal, river and lake patrols, management of the offices and liaison with the fishers. The sea-going surveillance personnel should be recruited from fishing communities, where appropriate personnel can be found. The air surveillance personnel can possibly be seconded from the military, or possibly local airlines to minimize training requirements. Support personnel will also be required from local staffing pools. Maintenance personnel for equipment will need to have experience, or be trained, on the equipment provided. The training for fisheries technicians and officers is key to ensuring competent staff for the implementation of fisheries plans. This topic is addressed later in this publication in the annexes.

4.2.5 Financial considerations

There can be a tendency when considering a new system to look for the most advanced technology which can do the work required. This is usually also the most expensive equipment on the market. Keeping in mind the economic logic that the overall expense of a conservation system should not exceed the benefits gained from the fishery, then it is more prudent, especially in the current economic climate, to look for *appropriate and affordable technology for each fishery situation.*

Cost effective data collection and verification

The requirements for *data collection*, and verification, usually include information on the fishers, their fishing vessels and gear, their individual and community dependence and returns from the fishery and hence, the period, quantity and value of their landings. Further information on the area of capture and the size, weight and age of the fish are also beneficial for fish stock assessment modelling and stock predictions, despite their inaccuracies. It could follow then that the most cost effective strategy would be to have data collectors in each, or many, of the large fishing communities. It could also follow, that any strategy which falls outside these basic parameters for data collection should receive their funding from other than government sources, or, if these are at the request of the fishing industry, then the industry should bear the cost, especially if can afford to do so.

The initial data collection task is a complete census of fishers. One cost effective strategy suggested is to use inter-agency resources to collect this data when the regular census of the population is taken. This can assist in providing basic, and later, updated information where this is needed.

The key role of the data collector, following the establishment of the data base, is the verification of the fish caught and that fish which is landed with additional information on the area of capture, size, age and sex of the catch. The updating of the initial data base can often be accomplished during regular data collection activities. Some states use personnel on a part-time basis for both data collection and enforcement, but this has conflict of interest difficulties where fishers may believe that the only reason personnel are collecting data is to "catch them" in non-compliant activities. This can lead to falsification of the data provided. Many countries already have extension officers, and community development officers from one ministry or another and these personnel can provide a wealth of information if it is channelled, or accessible, to the fisheries database. Data collection reliability is often enhanced when the individual collecting the data is known and respected in the community. The data being collected are, after all, personal information on catches, fishing areas and income, all which can be interpreted as being sensitive to the fishers. The confidentiality of individual data must be assured throughout this exercise.

Coastal

One example of a low cost method to gain information, if the fisheries are close to the coast, has been used successfully in New Zealand and some of the countries in the South Pacific. This involves the deployment of part-time coast watchers at a minimal cost to the State and the price of binoculars and a radio. The concentration of fishing around the coast is easily determined using these part-time personnel.

The key to ensuring quality information from the domestic fishers is to maintain credibility and close contact with them in their communities. This contact will also assist when seeking their ideas and support for new management measures.

Offshore

There are several options for the offshore portion of fisheries MCS. These depend on the value of the fisheries to both domestic and international fishers. Under the Convention on the Law of the Sea, the coastal State may establish the terms of fishing resources surplus to its current harvesting capacity. This includes a potential resource rent for the opportunity of fishing in the zone. This rent must be reasonable, but it can also be used to offset some of the costs of enforcement.

Port options

There may be several options to survey offshore fisheries activities. For example, it may be safer and more cost-effective to conduct inspections and transhipment of fish in port rather than at sea. Where possible, maximum use of port surveillance and inspection activities are encouraged as these can be very cost-effective and cheaper than at-sea inspections. It must be recognized that at-sea surveilance will still be a necessary component of most fisheries regimes.

Fisheries management strategy

One theoretical scenario of cost effectiveness could be a fisheries management scheme based strictly on *effort and area controls*, instead of quotas by species which wander all over the zone. The effort limitations would be established from catch rates in the past applied to each fishery. These could be verified through information provideed on the landing of the fish. In this case, the surveillance aspect would be greatly facilitated by being based on vessel sightings and effort monitoring instead of hands-on inspection of the vessels at sea to determine their catches. This suggestion, although ideal for surveillance, may be a bit impractical in terms of optimum utilization of the resources, e.g. combined fisheries in one area where the total for one species has been taken, but not that for the second fishery. It could possibly be an option for a single precies fishery. The idea has been presented to suggest innovative thinking for fisheries management options.

Using the principle of least cost to the State, however, if the fishing industry wishes to control the fishery by quota management instead of effort control, they might be encouraged, as part of the resource rent, to shoulder, or share the additional costs which may be incurred with such a system. Perhaps a combination of these principles could be effective if the fishing partner were willing to assume additional costs for surveillance to offset air coverage, observers or at-sea inspection costs for management schemes which they prefer tp have implemented. This inter-active management style would require a high degree of input and the acceptance of responsibility by the fishers for the conservation of the marine resources. In many cases it would require a complete change in attitude, consequently, it is a longer term strategy which one can consider for implementation with education of the next generation of fishers.

On the issue of *quota controls*, these have been found to be effective only if there is a timely acquisition of accurate catch data, including discarded and dumped fish that has been removed from the resource base. The acquisition of these data is also helpful for stock assessment, but it requires at-sea observations and inspections for quality control of the data being collected. This latter requirement is an expensive undertaking. It is, however, one rationale for fishers to assume the total costs of observer coverage, if quota control is the strategy they support for fisheries management.

A significant cost to the State is the establishment of a data collection system and analytical capability. There are several examples of fisheries information systems available, with cost effective regional systems being used in the South Pacific through the Pacific Forum network and in the Caribbean, through the new CFRAMP system. Several developed countries and international agencies can also provide ample examples and assistance in the custom development of appropriate systems.

Controls

Legislation

Fisheries legislation forms a major component of the *control* aspect of fisheries MCS. The fisheries management plan is transferred from theoretical ideas to legal requirements which form the base for the MCS operations, through the drafting and passage of fisheries law. It is at this juncture that fisheries managers, MCS officials and lawyers can assess the enforceability and cost of their management schemes. Common concerns expressed by MCS personnel all over the world are that the written law is often untimely, it no longer reflects the full intent of the management plan, it is overly complex, unenforceable and consequently, expensive to attempt to implement. The latter impacts on the credibility of fisheries officials in the eyes of their clients, the fishers. Cost effectiveness can be enhanced for this component if there is coordination and cooperation between the fisheries administration and legal drafters. This is an investment in time and greatly facilitates implementation of the final product for the benefit of the State. It also serves to strengthen the knowledge and capability of the legal authorities in fisheries resource management, a potential benefit to the development of other resource management legislation. It falls on the fisheries administration to foster this linkage with the justice department.

The internal policy decisions in the control phase of fisheries management relate to the strategy for implementation of the MCS operations and will be different for each fishery and state.

Licenses

One of the most important tools of fisheries management, which is often overlooked, is the privilege to fish, the fisher's license. This document is one of the most powerful documents in fisheries MCS, for it provides the fishers with the privilege to harvest an important resource, but it also sets the terms and conditions under which they may do so. The license can be used to provide all the base data regarding fisheries activities in the zone. It can require appropriate reports on fishing gear, activities in terms of time, location and catches, and can also require cooperation in fisheries management objectives. Further, it is the main tool which will serve to obtain the resource rent for the privilege of fishing in the State's waters. It can be the tool which establishes controls on an otherwise uncontrolled, open access fishery to meet the State's obligations under the Convention while minimizing the cost to the resource owners, the taxpayers. It can also be used to offset funding requirements for additional surveillance measures. The supporting legislation which makes the license so important is the legal right of the State to grant or remove this privilege and to exact sanctions for non-compliance. The license therefore contributes to the monitoring component of MCS through the requirement for base, and operational, data. It is the legal instrument for the control of the fishery and is key to the implementation of fisheries management plans.

Resource rent

The principle of "keeping it simple" is a concept to keep in the foreground in the surveillance component of MCS. The current times of fiscal restraint make this even more applicable today. Using this line of thought, coupled with the idea that the fishers should assist in funding any diversion from bare necessities for fisheries management, the following are examples of strategies which may be considered. Assuming that fisheries management plans have been developed and legislated in a cost effective manner, the Fisheries Administrator must then look at the human and infrastructure resources necessary to conduct fisheries surveillance operations. Assuming that resource projections have been determined based on one of the many stock assessment models (maximum sustainable yield, optimum sustainable yield, straight economic need, protein need, or otherwise), and that the basic requirement of the State is to ensure that the removals from the resources do not exceed these agreed totals, many options exist for enforcement. If there is only a requirement to report and land all fish captured, the surveillance function can possibly be carried out mainly with shore monitoring and spot checks at sea. It may also be the best strategy to use in a management system where the coastal resources are to be managed by the municipalities or sub-regions of the country, such as in Germany or the Philippines. Countries where fishing zones are very small and coastlines are short (for example, Togo - 48 km or the Congo - 41 km) might be States where the economics of the fishery would be best served by such a policy, if landing facilities are available.

Other zones, where there are large offshore fisheries or extensive coastlines, may require a strategy which depends on offshore technology. Some examples include the offshore component of the Philippines, the extensive zone of FFA, Australia, UK, New Zealand, Namibia, Morocco and Angola, as well as the island states of the Indian Ocean and the Caribbean Basin. Surveillance technology comes in various expensive packages including radar, aircraft, and satellite technology and cost effectiveness in these instances is very important. Successful examples include regional cooperation and sharing of expenses as seen in the FFA with the NIUE Treaty of 1992 and the subsequent, precedent-setting agreement between Tonga and Tuvalu for the sharing of MCS resources and the agreement for the flag vessels and resources of one State to assist in the enforcement of fisheries legislation in the other State's waters.

"No force" strategies

The use of cost effective "no force" tools is becoming most popular. These include the use of national or regional registry systems where the threat of removal of "good standing" is often enough to ensure compliance. Unfortunately, there are no international conventions in force concerning registration of ships, and none at all being considered for fishing vessels. The concept of international vessel registration standards and exchange of information would greatly facilitate the identification of vessels, implementation of flag State and port State control mechanisms and establish controls on "international renegade fishers" who are conducting fishing operations which undermine the internationally respected principles of conservation. The potential for the use and expansion of the regional register into an international principle merits serious consideration.

Another, "no force" initiative is the flag State responsibility for the activities of vessels flying its flag. This has been used effectively in the FFA treaty with the tuna fleet of the United States and is the principle behind the FAO "flagging agreement" and the efforts to control the fishing of highly migratory species and straddling stocks.

Another mechanism is the use of observers without enforcement powers, which, while being effective for data collection, has also been found to be a deterrent to non-compliant activities. This latter strategy is not without its pitfalls, the chief being the potential of external pressures on the observers at sea resulting in less than desirable data collection practices.

A new initiative is the development of vessel monitoring systems for timely catch and position information. The acceptance of this technology in the courts is one of the future challenges facing this technology.

Finally, a growing trend, borrowed from the commercial vessel trade, is the development and use of "port State controls" whereby there would be international agreements struck on a regional basis for the inspection and enforcement of fisheries legislation on any vessels operating in the entire region. This is an effective, low cost control using the potential of any country in the region being able to detain non-compliant vessels and crews as a counter-incentive to non-compliance with respected international maritime principles, be they for fisheries, pollution control or safety-at-sea.

These technologies can all be cost effective, and where they can be applied appropriately, they can be of little cost to the State other than the investment of time for coordination.

Traditional surveillance

Other traditional fisheries surveillance technologies include the use of aircraft and vessels to transport fisheries enforcement personnel to sea for inspection purposes. Air surveillance costs vary from $400 US per hour to $3,000 US per hour, or even up to $7000 US per air hour for large aircraft, depending on the aircraft and its capability. Offshore sea surveillance has a cost ranging from $500 per day to as high as $140,000 per day. These can be expensive resources for random patrol of fisheries waters. The capital cost of these acquisitions is often rapidly surpassed by the operational and maintenance costs, especially where trained technicians may not be available in the country. Careful consideration of all factors and needs is suggested prior to committing the State to the acquisition of major capital resources for MCS activities.

Fisheries management strategies

The fisheries management strategy, as noted before, will have a considerable impact on the MCS resources required. If the fisheries management plan are traditional and require at-sea patrols and inspections of the vessel, catches and gear, costs will escalate. These can perhaps be minimized by the legal requirement have on board fishing gear only for the fishery authorized for the particular area of operations. This can be verified, at least initially, through pre-fishing port visits which could serve for explanation of the legislation in force and the vessel inspection. This would not be an appropriate strategy for mixed fisheries with different size mesh. If the above were coupled with the requirement for all transhipment in port, this type of legislation and fisheries management scheme could be easier to enforce than other traditional methods.

The use of effort control as opposed to quota controls, which require expensive boardings, has already been noted. The use of air surveillance and photography linked with onboard navigation systems has proven expensive, but it appears to remain the most effective method to survey offshore zones for fishing effort control and area closures. In the case of the domestic fleet however, where transhipment at sea is not permitted, the most cost-effective means of verifying landings is through port monitoring as the fish is offloaded. This will not however, control the culling, dumping, fishing in closed areas or illegal transhipment at sea.

Regional cooperation

The advantages of cooperation with other states on a regional basis have been noted. The cost savings, facilitation of management of migratory stocks and the potential ability to fund newer technology on a joint basis are the main benefits that can accrue from such a relationship. Key to the decision on the infrastructure required is the complexity of teh fisheries, the trained resources available and the international cooperation possible in the region. The resultant costs should constitute an important component of the negotiation of fees for fishing privileges, both for the domestic and foreign fishing sectors.

Private sector MCS

An option gaining in popularity in some countries is to move the MCS costs away from the civilian and military bureaucracies to the private sector. Although the unit costs of MCS activities may increase initially above those for the government, it has been found that the private sector, with a clear mandate, can sometimes run the MCS operations in a more business-like, cost-effective manner than is possible in the government bureaucracy. Experience has shown this to be the case for air surveillance in some countries. The concept of community-based fisheries management and deputizing of private sector personnel to work with local authorities could also be a cost-effective option for consideration. The concept could be considered for other MCS components. Government officials then need only monitor the results of the implementation by the private sector. There is a major disadvantage here however, if the government does not remember to bring the MCS sector into the initial planning sessions to ensure that they can address and implement the fisheries management plans. Failure to do this can result in ineffective MCS activities. Privatization of MCS can be carried out successfully, but it needs a very good liaison between all involved parties, perhaps more than can be expected in some bureaucracies.

4.3   Management measures

Management measures are the specific elements of fisheries control which are embodied in regulations and which become a focus for surveillance activities. Cost-effectiveness needs to be considered for each management measure. The fisheries management plan, operational strategy and the management measures chosen for MCS must be included in the fisheries legislation. This will provide the base for implementation of the fisheries management plan.

One measure to be addressed is the use of *mesh size* for conservation purposes. It should be noted that this requirement can only be enforced in two methods; inspection prior to going to sea, with the provision that no other gear can be carried on board for that trip; or, inspection at sea which can only provide a snap shot of the fishing operations while the officer is onboard the vessel. The fisher may wish to prosecute a second fishery where a different mesh size is authorized and the earlier noted requirement then becomes an inconvenience. In the case of towed nets, if the fisher is permitted to use strengthening ropes to keep nets together when full of fish, and top and bottom chafers to protect the nets, and then trawls through weeds on the way to the fishing grounds, can small fish really escape from the net? This is further complicated by the fact that many nets are constructed with diamond mesh which has a tendency to close when pressure is applied during towing through the water. Some countries are now requiring square mesh design which stays open during trawling, except under extreme pressure. One can then question the advantages of using mesh size as a conservation tool in the case of towed gear, noting the difficulty of ensuring that it works properly.

The case for gill nets and entangling nets is different and the mesh size can be a significant conservation factor, but in these fisheries there will be little need to carry more than one type, or size of net, to sea; consequently, it will not inconvenience the fisher significantly to have the vessel inspected prior to fishing and ensure only one size net goes to sea for the trip.

The method for measuring a net should be standard, at least by country, and acceptable to the courts where fisheries cases go to court. Assistance of the judiciary to establish these standards would be advisable. Common standards include:

a.   measurement with a standard, graduated wedge or an implement with a standard width,
b.   measurement of the net when wet as it would be while fishing,
c.   measurement of the mesh stretched between opposite corners,
d.   measurement of several (a minimum should be set) adjacent meshes and averaging the result,
e.   measurement in the middle of the net away from any strengthening ropes.

A second management measure is the use of *chafers and strengthening ropes*. Chafers are attachments to the bottom and top of towed fishing gear to prolong the life of the expensive nets by reducing wear and tear from rubbing on the seabed. These can be made of netting, for top side chafers, or leather strips, for bottom chafers. These attachments are common and necessary fixtures to the gear, but a supplementary result is that they block the mesh and therefore retain all fish caught in the net. It has been found that the method of attachment of this gear does have an impact on the possibility of fish escaping the net. The bottom chafers are usually heavy twine or pieces of leather, especially for bottom trawls where the net is dragging on the seabed. These should only be attached at the tail of the net (the cod end). Topside chafers come in many different shapes and sizes, the main criteria normally established for this attachment being that it is attached only to the cod end of the net and in such a manner that it does not overlap the mesh to restrict the normal openings of the mesh in the cod end. Examples of two chafers are seen in the following figures.

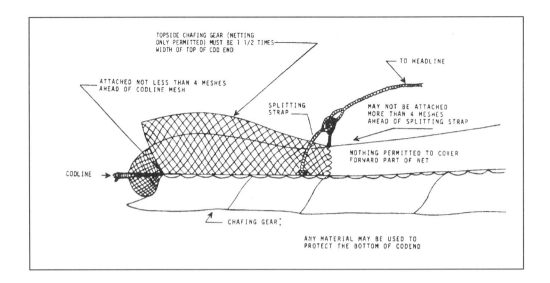

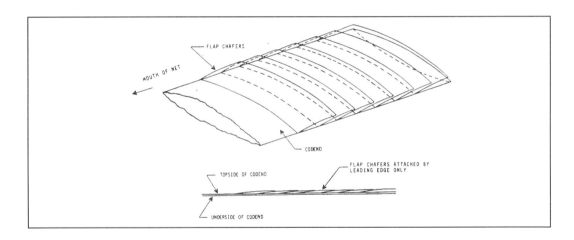

Strengthening ropes are necessary attachments to hold the net together and prevent it from ripping open when it is hauled on deck with a full load of fish. These are ropes which should be attached along the main axis of the net and where attached across the net must be attached in a manner as to ensure they do not reduce the size of the meshes in the net. The following examples illustrates this point.

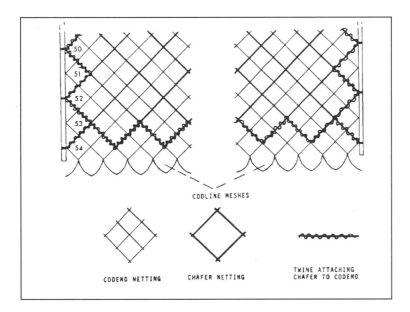

Strengthening ropes and chafers can only be checked during inspections in port and at sea. On larger vessels operating for days at sea, these can be re-laced to the net at sea and therefore become a management measure which is difficult to control without at-sea inspections or continual monitoring during the fishing operations. In the case of smaller vessels working inshore, port inspection would normally be sufficient as it is impractical to make changes during such a short trip.

Area closures are a management tool recommended to protect spawning areas during known seasons. This tool is also used to separate types of fishing gear whereby one type of gear is permitted, such as set gear, and another is prohibited, such as mobile towed gears. One lesson from the experience of area closures is that closures based on depth of water are unenforceable. More effective is the latitude and longitude designation of the area for such controls. This tool does not require at-sea boardings and inspections unless there is a prohibition of different types of stationary gear, such as traps and nets. The surveillance of area closures can be accomplished by properly equipped aircraft which have photographic and integrated navigational equipment and night flying and surveillance capability. This can minimize the necessity for sea patrols, but the presence of a patrol vessel is still the best deterrent. It has been noted that air patrols, although expensive, cover a vast area and are more cost effective in gaining necessary fisheries management and MCS information than sea patrols.

Some countries use *windows* as a tool to control fishing in their zone. In this case, areas for authorized fishing are established by latitude and longitude and all other areas are closed to all fishing. These windows can be set for different gears and patrolled by aircraft. Windows or area closures can be used effectively to reduce gear conflicts, especially between offshore commercial and artisanal fishers. On some occasions, countries have set area and time closures to permit different gear types to use the same area, but at different times. Another advantage of windows as a surveillance tool is that sea patrols can then be concentrated on these areas for at-sea inspection for compliance with regulations which are not enforceable by any other means. The disadvantage of windows is the impact this management strategy may have on the fishers and their reduced ability to chase the fish over their migratory paths. Windows can restrict fishing to the point that fishing opportunities are not viable. It is best to ensure that such a strategy does not overly impede the chance for fishers to access the resource. Stocks with low migratory patterns could be considered for this management measure.

There are various types of *catch or quota controls* used by governments today. These can be daily, seasonal or trip catches, zonal quotas, vessel quotas and annual quotas. In each case, there is a requirement to be able to verify, on a timely basis, the actual catches of fish by each vessel, species and possibly by area. This becomes a very complex administrative programme which is expensive in terms of finances and human resources. The monitoring and at-sea inspection and enforcement requirements to implement such a system are significant in terms of communication costs to support the necessary processing and analysis of data and verification of data quality. It is commonly accepted that control of the removal of resources from the sea is the most desirable conservation measure. Whether this is done through controls which count each fish or through less complex and less expensive methods, such as control of fishing effort, is a point which can significantly impact on the cost-effectiveness of fisheries management and MCS operations. It is accepted that where estimates of stock abundance are accurate, then catch controls would maximize the benefits to the fishers by permitting them the maximum removals, but this is an exercise which, to date, has been found to be very costly.

Trip limits are sometimes used as a measure of fishing control. These limits could include total catches permitted per trip or, more commonly, effort limits. The former requires someone to meet the vessel upon arrival to be present during the weighout of the catch. If a fisheries official is employed to monitor landings for the data system, this strategy would blend in with normal operations. If not, then it would necessitate extra personnel and effort. The disadvantage of trip, or catch, limitations is the temptation to discard all the lesser value, or small fish, prior to landing, thus possibly promoting the dumping of fish. The effort control requires reports of departure and return and a capability to verify these periods through sightings at sea. This is not impossible however, and will be discussed further as a viable, cost effective control mechanism.

A possible alternative to catch and quota controls is *effort control*. Where stock assessments are not very precise, there can be a measure of fish removals through effort controls by limiting the fishing time of vessels and fishers in certain areas. This can be implemented by assessing the experience of fishing units and then, based on a conservative assessment of the fishing efficiency of each fishing unit, limiting fishing to the time expected to take a certain amount of fish, with an appropriate safety factor for conservation. This can then be cost effectively checked through air surveillance and verified by port inspections prior to, and upon completion of, fishing in the zone, for the large offshore vessels. Coastal patrols and landing checks are usually sufficient for the artisanal fisheries. The at-sea resources then required can be minimized, thus making the system more cost effective. This is a system which can be implemented to establish some control of effort even where the exact resource status is not known, but where catches appear to be consistent in terms of quantity and size of fish. This would be a temporary measure until appropriate data gathering could be established and data analyzed for more accurate stock assessments. There would also be the need to seek the support of the fishers for information on the current fishery and to gain their support for such controls as a temporary measure where the situation may appear stable, but is actually vulnerable, due to the lack of data.

A growing trend in many countries is to use *individual transferable quotas* (ITQs) as a management tool. This system is employed for those fisheries where there is limited access to the resource, the numbers and identification of each fisher and gear are known, recorded and have a history. The system then provides for an allocation of fish to each fishing enterprise, or fisher. This is often by species and stock area for a specific period of time. The fisher then has the right to plan the fishery of these allocations during the period to maximize economic benefits. The fisher also has the right to transfer, or sell, the allocation, or portion thereof, to another fisher on a temporary basis. This initiative enhances the benefits to the fishers with respect to their cost effective harvesting and processing of the resource to maximize their economic returns.

The complexity of the system and the requirement for an advanced communications and data network to effectively manage this strategy makes it difficult for developing countries. Experience has demonstrated that this strategy is only appropriate for small island States with small fleets and few fishers, fishing for essentially an export market. These prerequisites are necessary for States to be able to implement the appropriate controls to successfully use this management strategy. It is for these reasons that the ITQ strategy is not recommended for developing countries at this stage of its evolution.

The ITQ system does provide an opportunity for success if there are available resources to successfully use new technology. Such an initiative is being attempted by the FFA at the moment and the successful implementation and enforcement of the vessel monitoring system (VMS) could make this an ideal management tool for the future. The key to the success of the ITQ system could be the VMS system if its eventual cost to countries and fishers is low.

Another management tool is the establishment of *minimum or maximum fish sizes*. The regulation normally specifies that the capture or landing of a fish of a certain size is not permitted. The intent behind the minimal size of fish is to prevent the harvesting of non-commercial sized, juvenile fish of little market value. This is then expected to assist in stock enhancement, maximizing future benefits to the fishers. The prohibition against large fish is usually intended to preserve the brood stock. Unfortunately, neither of these regulations can be enforced without continual monitoring at sea. The prohibition against landing can easily be subverted by culling the fish and dumping the prohibited sizes. Once the fish is caught, it is usually dead. Dumped fish are not normally recorded as catch; thus the estimates of removals from the stocks are therefore undermined. It is suggested that fishers be encouraged to land all their catches, that these be analyzed against their areas of fishing and if areas need to be closed to protect fisheries, then this be done. This implies that the use of fish sizing would not be a legislated management tool, but an indicator for fisheries managers to close an area where small fish, or large brood stock, are being caught. This suggestion might bear further consideration, especially where fishers are fishing in one area where all fish must be retained and also in another where possession of under-sized fish is an infraction.

The prohibition of certain fish sizes, although it may encourage the fishers to shift fishing zones, also encourages dumping and misreporting. The surveillance costs to enforce these regulations are difficult to justify unless there is a physical presence on the vessel at all times, or a high level of spot checks at sea.

Most states implementing a MCS system employ some sort of *vessel movement controls*. These are usually in the form of report requirements from the offshore foreign vessels and trip intention reports from the larger domestic vessels which are fishing longer than two or three days at sea. The vessel movement reports for the foreign vessels range from zone entry and exit reports, to port entry and exit reports and area changes.

In the case of domestic vessels, the port departure/entry and area change reports provide the basic movement information for MCS purposes.

All movement reports require the vessel identification which includes the vessel name, call sign and the master's name as well as the fish onboard by species and intended activity. In the case of the foreign vessels, the fish onboard and intended activity are important for surveillance and catch monitoring in the zone. If the vessel has already picked up its fishing license at an earlier date, then the MCS authorities need to decide whether an inspection is necessary and if this will be at sea or if the vessel will be ordered to port. This first zone entry report for foreign vessels draws up the information for the vessel and commences the monitoring exercise, which will continue until the vessel has departed the zone and all reports and documentation have been received.

The zone exit report, which is commonly required by the central fisheries control, before departure from the zone, reiterates the vessel identification information, and the time and position of expected departure from the zone. This report is the final opportunity for the Fisheries Administrator to inspect the vessel by intercepting it at sea or ordering it to port. The matter of permitting multiple exits and entries to the zone for fishing purposes is one with which Fisheries Administrators will have to contend. A vessel could have good reason to exit and enter the zone due to medical reasons or such, or it could be for fishing operations outside the zone, transhipping its catch, changing fishing crews and thus avoiding coastal State regulatory requirements. As all the latter reasons impact on the fishing efficiency of the vessel, the coastal State may wish to include in its legislation and fishing agreements, provisions to capture the information required for conservation purposes. If coastal State fisheries officials are aboard the vessel upon departure from the zone, their authority for fisheries surveillance and enforcement may be challenged. It is best in these sensitive situations to decide early whether the vessel will be permitted to depart the zone; if not, the order to "halt" must be given prior to departure from the zone to establish the parameters for "hot pursuit" (Article 111, Convention on Law of the Sea). The zone exit report thus triggers the action for decisions regarding the control of the vessel prior to its departure from the zone.

Vessel sightings reports are collected to update the fishing vessel monitoring information database. These reports are usually standardized and are completed by all MCS resources and sent to the MCS control centre for collating and updating of current information. The report normally includes vessel identification information, such as the name, vessel marking/call sign, home port and information as to the activity of the vessel. If the vessel is steaming, the report includes its course and speed, if these can be determined. If it is fishing, the course, speed and type of gear it is using are necessary. If photographs can be taken of the fishing activity, including the gear, e.g., lines in the water, etc. these should be taken and labelled with the vessel identification information, position and time of the pictures. The vessel sighting report should also record that photographs were taken, and the numbers on the film. This could all be necessary information for the courts if it were later found that the vessel was not fishing in compliance with its license, or fishing without a license. These are fairly common standard procedures which all MCS personnel should be trained to do on each sighting. Sightings form the key verification of the vessel fishing effort in the zone and can also be used to estimate catches.

Vessel inspections are a key management tool for monitoring and surveillance. Port inspections are considerably easier to conduct than those at sea due to the safety factor of not having to deal with the motion of the sea either on boarding and disembarking, or during the inspection itself. The detail of the vessel inspection, either at-sea or in port, depends on the Fisheries Administrator. Naturally, it will be impossible to see the fishing and processing operations during an in-port inspection, but it should be possible to reconstruct the fishing activities of the vessel since its entry into the fisheries waters of the State. In both cases, one should be able to determine the fishing pattern, catches, and verify the fish onboard the vessel through an inspection. It should also be possible to check the storage and size of the fishing gear, at least that on deck, for compliance with fisheries legislation. The fish onboard the vessel can be determined as precisely as desired if the funds for off loading and reloading are available. This is not normally recommended unless there is sufficient reason to believe there has been a violation. The advantage of at-sea inspections is the monitoring of the handling of the gear, the processing and storage, and verification of the processes for handling waste fish. This can provide information for fisheries management planning and possibly catch and effort information which would be useful in access negotiations with foreign partners. Efficiencies of processing methods and equipment can be monitored and used to validate conversion factors from processed to round weights.

The accuracy of vessel inspections, both in port and at sea, is a crucial component of the surveillance aspect of MCS. It is the initial inspection which verifies the fish onboard the vessel and forms the base data from which final assessment of fish caught in the zone can be determined. The intermediate inspections during the authorized fishing period provide verification of compliance with fisheries legislation and obtain data used in the determination of catch rates and catching efficiency of the fishing unit, processing efficiency and handling of waste product. This is all information which permits analysts to determine with a certain degree of accuracy, the real, round weight removals from the fisheries resources.

The use of fisheries *observers* is a management strategy that may not be appropriate for all countries. A large measure of the success of observers depends on their professional competence, their personal integrity and, in short, their honesty. An observer programme, to be effective, needs close supervision and appropriate checks and balances to validate the accuracy of the data collected. It can be very difficult to entertain the implementation of an observer programme without a counter-incentive to external financial incentives to turn a blind eye to existing practices. This can be true not only for observers, but all enforcement activities. Some countries have implemented a sharing of proceeds from successful prosecution of offenses. The concern here then becomes the resultant vigilante attitude which sometimes grows with this policy. This can be especially concerning if there is a proclivity for levying administrative penalties without appropriate checks and balances on the validity of the information respecting the alleged violation. Fishing masters may plead guilty to get out of the country and fishing waters, rather than defend themselves in a long drawn out case. The resultant effect eventually leads to reduced fishing in the waters and reduced cash flow from these fishing units.

Fisheries observers, where this management tool can be utilized, can be one of the most cost effective MCS schemes available to the State. There have been several attempts at implementing observer schemes for fisheries MCS. Some of these have achieved less than expected success for several reasons, including the following:

1. Observers have been employed in a dual role of observer/crew member, thus compromising their role and loyalty to their fisheries duties.

2. Observers have been paid directly by the vessel master, thus further complicating their conflict of interest position between fisheries and the vessel master.

3. Observers have been remunerated at such a low level that they are very susceptible to extraneous financial incentives to "turn a blind eye" to certain fisheries practices.

4. Observers have not been appropriately supervised or trained to execute their duties effectively.

5. Observers have been given, or perceived themselves as having, enforcement powers, thus complicating their monitoring role and serving to make their reception on board the vessel difficult.

Success in observer programmes seems to have resulted when:

a) observer services have been funded by the fishing enterprises through the government, or some other coordinating body, such as the private sector;

b) the observers have been employed solely as observers, that is, monitors for the government (also in gathering much scientific and technical data) and advisors to the vessel masters regarding authorized fisheries activities;

c) observers have not been granted enforcement powers;

d) appropriate training and evaluation have been conducted for observers; and

e) observers have been paid appropriately to negate the incentive of bribes.

f) observers can effectively watch for compliance with regulations and report possible violations to enforcement authorities, who subsequently endorse.

One of the key management tools available to Fisheries Administrators is the *fishing license*. This document establishes the legal rights, privileges and obligations of fishers. In the past, countries have set the royalty for fishing on the quota or catches of the vessel. This is more common in the case of international fishing partners than domestic vessels. An unfortunate consequence of this practise is the incentive to misreport catches to avoid license fees. The more successful fee strategy is one based on a fee per vessel with no basis on catches. This could also become a basic nominal fee, but it is suggested that the larger vessels with greater catching capacity should pay higher fees for the license. The need to verify catches for license fees and the incentive to misreport and falsify records is removed by these strategies. The license system, as noted earlier, provides the base data for MCS activities and fisheries management planning. Its use has often been concentrated on international fishers, but noting the extent and impact of national fishers on the resource, it can be a very effective control mechanism for all fishers.

The use of *new technology* is always an attractive option. There are new satellite capabilities, vessel monitoring systems, new aircraft, radars, infrared equipment and photographic technology and vessels, all which can be attractive to MCS officials. The only caution which can be offered is for Fisheries Administrators to look carefully at the use of the technology in local operating conditions and make the assessment and decision based on these results, with appropriate attention to the cost of procurement and, more important, operations and maintenance costs and the capability to carry these out in the country. New suppliers, if they wish to sell this equipment, should be willing to fund trials in the local environment if they are serious regarding the efficiency and effectiveness of their equipment for the task to be done. It is the supplier who should shoulder the financial responsibility of assessing the appropriateness of the equipment to the situation before attempting to sell this equipment. It remains the responsibility of the Fisheries Administrator to assess the performance. The operations and maintenance training for the equipment should be included in the cost of the equipment and provided by the supplier, if it is accepted.

There has been a growing trend, due to the relatively high cost of MCS operations, to seek "*no force*" MCS strategies. The intent here is to pick those MCS tools which can exercise sufficient monitoring and surveillance controls over the fisheries resources to meet government needs at the lowest cost possible. The most popular "no force" strategies seem to be the use of a national, or regional register of vessels of *good standing*. Only vessels on this list would be eligible for licenses to fish in the fisheries waters of concern. Information included in the register includes the identification of the vessel and master and a record of performance and compliance in the fishing sector. In a regional situation, information is shared between parties and decisions can then be made regarding a vessel and master in the entire area. Vessel masters and fishing companies have, over the years of experience in the FFA, been seen to respect this tool, to the point that potential removal of *good standing* has been enough to ensure compliance or appropriate action in cases of a difference of opinion regarding fisheries MCS activities.

As noted earlier, vessel registration is an area which merits considerable attention in the future for both fisheries control/MCS purposes and also vessel safety requirements which impact on safety-at-sea and protection of the fisheries habitat. At present there are no international conventions or standards for registration of fishing vessels. The potential exists for this to be a credible international management tool for both flag State and port State control mechanisms which could benefit developing countries with information on third party vessels applying for registration in their countries.

Other tools in this category of "no force" mechanisms include the port inspections, flag State responsibility for the actions of its vessels in the zone and observer programmes. Many countries have required a representative from the international fleets which are authorized to fish in their waters to be resident in the State and to be responsible for the actions of the fleet in these waters. This has been found to be effective only if the representative has the appropriate authority over the vessels in the fishing fleet. New tools in this category could include the idea of effort control, vessel monitoring systems and others which would require minimal effort and expenditure from the coastal State for use of its MCS resources.

4.4 Fisheries law

The development of fisheries management plans is, ideally, the result of analyses of biological, social and economic information and appropriate formulation of fisheries management strategies. However, fisheries management needs a legal base from which to implement fisheries operations for the conservative management of fish resources.

4.4.1 National obligations

Convention on Law of the Sea

The legal instruments which form this key legislative base include the fisheries acts and regulations developed pursuant to the planning exercises. These legal documents must, of necessity, for national and international credibility, take into consideration the terms of the Convention on the Law of the Sea, and any bilateral or multilateral treaties or agreements in effect regarding fisheries in the zone of influence and control of a given country.

There are several articles which have great impact on fisheries management in the Convention and these are summarized in Annex I. At the risk of over-simplification, there are several articles under the Convention which attract the attention of each coastal State. There is the definition of the three zones of interest: the territorial seas (most commonly twelve nautical miles from baselines established under various agreed rules around the coast); the contiguous zone (which commonly extends 12 nautical miles beyond the breadth of the territorial seas, e.g., twenty-four miles from the baselines), and the third zone (the exclusive economic zone) which is usually not more than 200 nautical miles seaward of the noted baselines. Some states have extended their territorial seas out to 200 nautical miles, but these are rare and are not consistent with the Convention on the Law of the Sea. A comprehensive listing of limits of territorial seas, fishing zones and exclusive economic zones is included in reference 33, Coastal State requirements for foreign fishing, FAO 21, rev. 4.

International agreements

It must be recognized that both bilateral and multilateral agreements can have a significant effect on fisheries management and MCS systems and strategies. Negotiation for access to fishing grounds of developing countries can be an interesting and very challenging exercise. The coastal State is looking for financial benefits from the negotiations, whereas the international party is looking for access to the fishing zone. This usually results in compromise which in effect will undoubtedly impact on the MCS strategy and the legislation for fisheries management implementation. The fact that it is usually a distant water fishing nation which is seeking fisheries access, possibly having greater experience in the process, with an established pattern for negotiating benefits weighed heavily on their side, can place the developing country in a very disadvantageous position. The coastal State's requirement for foreign currency can also place undue pressures on the negotiating team in favour of the negotiating strategy of the foreign party.

A review of past agreements of distant water fishing nations has demonstrated the advantages to these nations, sometimes at the expense of the developing countries.[9] Negotiated international legal agreements take precedence over existing legislation and consequently, fishing vessel licenses for these fleets sometimes have clauses which give them a fishing advantage over other fishers. This is not to say that international fisheries agreements are not beneficial to coastal States, as many countries have received considerable assistance through these processes. Whether states have received full value for the resources they have negotiated away is a matter for conjecture. An example of such negotiations concerns the EEC, especially since expansion of the fishing fleet with Spain and Portugal joining the Community, whereby their aid package was very much linked to fisheries negotiations. For several years, the EEC provided fisheries aid and training for all fisheries components except MCS activities, and there was no explicit recognition of the rights of artisanal fishermen in their agreements. Recent agreements with the Seychelles and Madagascar have been more comprehensive (regarding inspections prior to departure from the zone) but the developing State often does not have the capability and infrastructure to implement the letter of the agreement.

Other distant water fishing nations have refused to recognize regional fisheries agencies and have insisted on bilateral negotiations. This tactic has now been all but broken by the FFA, through the internal regional agreement to insist on minimum terms and conditions for international fisheries agreements. This accomplishment is another achievement of regional cooperation.

4.4.2 Impact of *civil* and *common* law systems

One of the common concerns of fisheries personnel lies in the final output of litigation processes. Fisheries Administrators have expressed their concern over the apparent lack of success experienced in this area of MCS operations. There is definitely the need for training in this activity to ensure that evidence is gathered correctly and presented in such a manner to result in successful prosecution of fisheries cases. Another influential factor is the legal system of the States.

The system to gain a conviction for an offence in the civi law states differs from the system in the common law states. While in the civil law system, the administrative procedure for imposing a sanction provides for an appeal before the Minister, this is not always the case in the common law system. In relation to evidence under the common law system, certain technical rules, such as the rule against heresay, restrict the range of evidentiary material that can be introduced. This can cause difficulties for the introduction of certain types of

---

[9] *Sevaly Sen, (1989) noted the similarity of negotiated agreements from the EEC and their potential negative impact and disregard for artisanal fisheries.*

information in the courts, such as that obtained from vessel monitoring systems and legislation may be necessary to accommodate this type of information. On the other hand, in civil law states, a report from an officer (inspector) is accepted as *primae facie* evidence. Without going into significant detail on the differences between the two systems, this provides and example of the impact of the legal system on the implementation of MCS strategies.

4.4.3 Core components of fisheries legislation

Fisheries legislation is usually comprised of two main instruments;

1.   The primary instruments are the Acts which defines the general parameters, authorities, powers, requirements of the fishery, including the infractions and penalties for each.

2.   The second instruments are the Regulations. The regulations provide much greater detail on the technical aspects of the fisheries, permissions, responsibilities, obligations and infractions of fisheries law.

The Act is a more permanent document and normally requires full Cabinet or Government approval for revision. The regulations, seen as the every day instruments for managing the fishery are usually much easier to amend, either through Cabinet or sometimes Ministerial approval. In either case, it is recommended that the entire law be reviewed regularly to ensure that it reflects the fisheries policies of the government. It is awkward for Fisheries Administrators to lay charges for a fisheries offense and then find that the infraction is one against policy only and the appropriate law to implement the policy has not yet been proclaimed. It should be noted that a review of such cases has revealed occasions where administrative penalties are permitted as part of the legal procedure, to misuse these to secure a guilty plea for an infraction when that infraction is not yet law.

There are several core components which comprise most fisheries laws. The introductory section usually includes a section on **definitions**. Key definitions address the Minister responsible for fisheries, the Fisheries Administrator, fisheries officers and observers, if necessary. Other definitions which may be linked to other legislation would note the various zones and their limits in the law. Finally, there are definitions of fishers, licences, the act of fishing and vessels which may become especially important in common law courts when prosecuting an offence.

Early in the legislation there is a section on **authorities**. Fisheries management needs a clarification of the authorities of the ministry responsible for fisheries and its powers over the legislation and implementation of these laws in the offshore, coastal, riverine and lake waters. The authority to present and amend the laws and the scope of the law is usually clarified in this section. This section can also clarify the relationship of other ministries to fisheries on MCS and fisheries matters.

The authority of the Minister is required with respect to powers to enact legislation through regulations to implement fisheries management plans. Here the civil and common laws differ, in civil law the Director General for Fisheries has the authority to issue licenses for fishing activities, cancel or suspend said licenses, appoint persons to positions of responsibility in the Ministry, and to exact administrative penalties for fisheries offenses. In the common law system this authority rests with the Minister, who may delegate this authority. In the case of civil law there is an appeal tribunal usually available to the accused where a serious offence has been committed while in the common law system this option is not always present.

The specific authorities and powers of each key member of the fisheries department, management and field staff, are normally amplified in this section with respect to management authority and punitive responsibility and powers, monitoring and data collection requirements, sampling, inspection, search, seizure, detention, prosecution and confiscation procedures. Further, the law usually formally states the appropriate reception expected for fisheries personnel and observers or others acting under the authority of the government. This latter point is very important for governments which may consider privatizing aspects of their fisheries MCS activities and for certain aspects of monitoring, such as dockside monitoring and observers.

The responsibilities, as well as the authorities, of each of the individual positions is also stated in the legislation. An example is the responsibility for fishers to accommodate and feed observers at the officer level of their vessels, but the law could also note the monitoring and advisory role of observers *excluding enforcement powers*.

The details of the fisheries management plans are outlined in the law and the **obligations of fishers** in this respect are then stated. The heavy pressures on most fishing stocks have resulted in the consideration of more stringent conservation measures including not only measures that regulate fishing activity, such as closed seasons and areas and gear types and uses and also, fishing effort. This latter point can change fisheries from open employment opportunities of last resort to a closed fisheries profession through the imposition of a limited entry fishery concept. The terms of entry, the requirements for licensing and criteria for application for such licenses can also form a part of the legislation. With respect to the use of licensing as one of the management tools, the two tiered licensing system (one license for the vessel and gear and another for each fisher) has been most popular and effective to date. This is opposed to licenses for each component, vessel, gear and fisher, or the combination of one license for the fisher which is further limited by a particular vessel and a specific gear type. The license application forms collect the basic information for the MCS database for both fisheries planning and MCS operations. Information on the fisher, vessel, gear, operations planned, social aspects and demography is also collected during this process.

The obligations and requirements for certain management measures are set under the fisher's licenses. They include the requirement for *marking of the fishing vessel* for ease of identification. This requirement has now been standardized throughout the world as a result of FAO efforts in 1989. Markings are clearly specified as to the size and location on the vessel, reflecting the size and type of vessel. These specifications are attached in Annex A. The vessel marking requirements are now recommended for inclusion in all fisheries legislation. This facilitates both the monitoring and surveillance aspects of MCS operations from the air and the sea. Linked with this vessel marking publication is FAO Technical Paper 267 on the Definition and classification of fishery type vessels. Examples are included in Annex A, but it is recommended that all fisheries personnel become familiar with this publication for surveillance recognition of fishing vessels which may operate in their zones.

An increasingly common requirement for fishers is to *mark their fishing gear*. This has been assisted by the standard definition and classification of fishing gear types (see FAO Technical Paper 222 rev. 1) which should be a document in the library of all Fisheries Administrators. Examples of standard fishing gears are noted in Annex B. The requirement to mark the fishing gear in a clear and obvious location will also facilitate the monitoring and surveillance aspects of MCS activities.

The matter of *logbooks* has been one of considerable discussion for several years. Logbooks serve many uses, first for information on the fishing operations of the vessel and, second, this information, if in sufficient detail, provides important input into the biological stock assessment component of fisheries management. It is these two aspects of logbooks which create discussion between fisheries managers and scientists, the latter seeking additional detail and knowledge. The MCS requirement for logbooks is for monitoring of the levels and the areas of the catches, which then contributes to future fisheries management planning and also serves as a verification of compliance with the terms of the fishing license. The biological component requires greater detail on the fishing operations which may include, in addition to the aforementioned, the depth of fishing for each set, or haul, length of the gear and/or time of trawling, species breakdown by set, or haul, size of fish and age and temperature of the waters during the set, or haul, at the depth of fishing. The economist would also like to know the processing methods, packaging and labelling, waste factors, and processing of fish offal. Other fishing operations such as transhipment of fish, quantities by species and location and names of the vessels involved in the transfer are also details MCS officials wish to record and analyze. These points are presented to note the complexity of the discussions regarding this topic in the past. This has lead in some situations to two or three logbooks for vessels for fishing, production, and transhipment. In addition to the requirement for this logbook information, it should be remembered that this is one component which is commonly standardized for regional cooperation, if such an initiative is being contemplated. It is obviously necessary to agree on some standards for the collection and recording of the data in a manner which is usable to fisheries staff. Most countries have found it effective to design their own logs and issue them, with instructions, to fishers. It has proven to be less than effective if information that is not needed is collected, and this should be avoided, regardless of the temptation to gather it.

Transhipment has been an issue of some importance to MCS operations. There are several philosophies regarding transhipment, ranging from no permission for transhipment inside the zone to permission for transhipment in the zone, but with prior notification to authorities, and finally, to permission for transhipment, but only in port. Fishing vessels normally resist transhipment in port, due to the bureaucracy involved, as they can be treated in the same manner as normal transport vessels with no recognition that they are in a special category, due to their cargo. When there is no transhipment allowed in the zone, this will inconvenience the fishers, but the impact is also a loss of data on catches taken in the waters of the State, as the vessels will tranship outside the zone where there is usually no requirement for the master of the vessel to permit an inspection. Verification of the catches in the zone will be lost, unless there is a requirement to report the fish onboard on entry and if exit from the zone is not permitted without an inspection. This becomes administratively difficult. Transhipment inside the zone is difficult to monitor and has its level of difficulty with respect to safety-at-sea. The recommended approach is the requirement for all transhipment of fish to be authorized for coastal State ports recognizing that although this may be expensive for the vessel, and hence impact on the revenues expected from the access agreement, it must be balanced against the risks of inaccurate, or inadequate, information on retained catch. The monitoring aspect of port transhipment is safer and much more accurate, thus being a benefit to MCS activities. It is for this reason that countries may wish to review their port State administrative requirements with a view to encourage fishing vessels to tranship their catches in port. The legislation must reflect the decision in this regard.

The matter of *reports* is key to the successful implementation of MCS strategies and input into fisheries management planning. There are several reports which Fisheries Administrators may wish to consider for their fishers. The requirement of a catch report is most common. This report is sometimes collected by port monitors verbally from domestic artisanal fishers. The fishers on larger vessels are often required by license, and regulations, to complete these reports on a daily basis and return them, at the end of each trip. If the trips extend past one or two weeks, it may be advantageous to receive a summary report of the catches for the week. These reports are usually for a set period and include at a minimum, the name of the vessel, side number, catches by species in round weight, broken down daily over the period of the week. The position of the vessel at a standard time each day is included with this report to give it geographic reference. The report serves as a document verifying the fishing effort of the fishers, the location and timing of this effort for fisheries planning. In addition, it can be used to ensure the vessel is complying with its license in terms of species being caught, effort permitted in the zone and location of the fishery.

If the management strategy is based on effort control using assumed daily catch rates, including a factor for culling, there would then only be a need for a report of the vessel identification and its positions of fishing over the period. The requirement for the return of completed logbooks and a port inspection prior to leaving the zone can be effective methods to verify retained landings, and possibly catches, for future effort control measures as required. This strategy would be strengthened by the legal obligation to tranship fish caught in the zone in local ports. There would then be a requirement for port inspectors, air surveillance and possibly fewer random sea inspections. The use of satellite technology with vessel monitoring equipment could also be effective in this case. The cost effectiveness of reduced sea inspections could be significant. It must be noted that this strategy would only have potential for single species fisheries and could not effectively accommodate multi-species and mixed species fisheries.

The requirement for periodic position reports from all fishing vessels during their fishing activities has become a standard practice. The rationale for this is twofold; one, for verification of compliance with legislation and licensing provisions, and two, for input into research programs on stock assessment. There are other key reports for surveillance purposes, including reports on entering and leaving the zone and, in some cases, when changing sub-zones within the fishing waters of a country, or possibly, when changing a fishery. The entry report, with vessel identification, its position in latitude and longitude on entry, with the breakdown of fish species on board, provides the first check of the vessel and the fish being brought into the waters. This can be verified through estimates if the vessel is required to visit the port to pick up its license and receive a briefing before commencing fishing operations.

Many countries require the vessel to report prior to departure from the zone and the intended departure point. This then permits planning for post harvesting inspections at sea if this is a management decision. Other management strategies follow the domestic principle whereby the vessels are required to come to port on completion of fishing in the zone. This latter serves many purposes. On the domestic side, certification of the fish caught and retained is easily obtained from weigh outs of the fish off loaded. This assumes there has been no transhipment of fish at sea, an operation difficult to monitor without regular air surveillance. With regard to international fisheries, the final port visit permits verification, through estimates, of the fish onboard and, taking into consideration authorized fish transhipment, will provide an estimate of the fish caught in the zone as a check on the reports from the vessel and later, the logbooks. Finally, many fisheries agreements require the off loading of a percentage of fish caught in local ports for domestic consumption or to support local fish processing infrastructure. This final port visit is the time to obtain this fish. It must be realized that DWFN vessels will often prefer to pay a penalty rather than off load fish as it is viewed as an expensive undertaking and an unnecessary tax by the government.

It has been previously noted, that there is often a requirement for fisheries personnel to be onboard fishing vessels. It has been found from past experience that government officials are not always regarded with favour when they visit the vessel, especially if it is for a prolonged period. This is particularly the case if the vessel is required to carry fisheries observers. It is for this reason that it is recommended that legislation note the conditions of treatment of officials expected by the State, for example, simply stating that all government officials are expected to receive living conditions at the status of the ship's own officers. On the other hand, government officials must remember their obligations in that they, and their vessels, should be properly identified and marked at all times. Further, they should conduct their activities in such a manner as to minimize the interference with the fishing operations of the vessel as this is a negative cost factor for the fisher.

Finally, the legislation, fisheries agreements and licenses may, due to the requirements stated above, necessitate the vessel and fishers to carry certain documents and equipment to meet the terms and conditions to fish in the waters of the State. If this is the case, it is best stated as a legal requirement in teh laws or regulations. One such example is the carriage of the licensing and report documents for verification by fisheries officials as requested. Earlier, it was stated that a fisheries management scheme based on quota management is very difficult to monitor accurately, without timely and accurate reports of catches by species and area and the capability to verify these data. The South Pacific Forum Fisheries Agency has made great strides in this direction with the pending implementation of an automated vessel monitoring system for the large offshore vessels operating in that area. Each vessel will be required to carry and operate an integrated satellite communications and Global Positioning System (GPS). This system will be able to be interrogated from the central regional fisheries surveillance centre. The position information will come automatically from the system, but the catch information will be entered into the system by the vessel. Future options may include methods, through integrated log sensors, to automate this aspect as well. The potential to tamper with the transponder is also an issue which will need to be addressed both in the design and the legislation for the application of this technology. The point is that the success of this initiative can revolutionize surveillance and reduce costs dramatically in the future, with port inspections and port State controls then assuming greater significance. The equipment for this monitoring and surveillance is a requirement by fisheries law and hence becomes a cost of fishing for the vessel operator, not the coastal State. Air surveillance, although still necessary to identify unlicensed fishers, can be more effectively targeted to radar targets which do not respond to identification queries from the transponders.

The legislation, usually reinforced through the tool of the fishing license and surveillance activities, states the *authorized and/or unauthorized fishing activities* for the State's fishing waters. Examples include the use of fishing gear by type, fishery and area, mesh sizes, chafers for gear, operating areas by species, periods of permitted fishing, transhipment, port entry and exit requirements and other such requirements. The law may state these authorities in a negative manner as well. These would then become unauthorized activities, examples being, trawling inside the artisanal fishing zone, running over other fishing gear, and setting fish attraction devices (FADS) too close to other gear.

A common final section in fisheries legislation, specified in the Act, is a clarification of the *infractions under the law and the penalties for each infraction*, or group of infractions. This section is very important to the success of fisheries MCS operations and the entire fisheries management process, for it is from this section that the full force of the fisheries process gains its credibility and its possible deterrent effect.

The Act will normally categorize infractions and penalties according to severity of the activity. Most legislation provides for automatic forfeiture of illegal gear and fish taken during the fishing operation on a finding of guilt for the offender. Seriousness of offence varies from state to state, but legislation usually includes among the offenses the following: assault on a fisheries official, unauthorized fishing, fishing contrary to license terms and conditions and infractions of the fishing regulations.

The penalties section has caused considerable discussion, especially with respect to the Convention on the Law of the Sea. The Convention prohibits automatic confiscation of a fishing vessel and gear on the alleged commission of an offense, but could be interpreted as permitting this penalty on conviction. More serious discussions have resulted from the prohibition of imprisonment for violations of the fisheries legislation within the EEZ (Article 73 para 3). Some coastal States have ignored this clause in their legislation and have indeed imprisoned both national and international fishers for violations of fisheries legislation. More common, however, is the quick release of the vessel and crew following submission of an appropriate bond to the government.

Further under the penalties section is the difference in the onus of proof under the civil and common law systems. The latter lays greater pressures on the prosecution to prove the details and guilt surrounding the offense, while the former accepts as fact reports from enforcement officers. There has been, on occasion, an attempt to place in legislation the "reverse onus of proof" principle, the concept that the accused is guilty until proven innocent, or the onus falls on this accused to prove innocence. There has been an interesting negative reaction of the judiciary resulting from the use of this concept in many common law courts. The judiciary tends to view the concept as a excuse by the prosecution for not being able to prove its case and possibly be contrary to one's express or implied legal rights. This rather interesting negative response places even greater pressure on legislation drafters to ensure the law is enforceable. In some countries, the acceptance of carefully drafted, *primae facie presumptions* has been effective in alleviating the difficulty experienced with "reverse onus".

The level of penalties is also another issue of some concern in developing States. It is common to forfeit any fish illegally caught and illegal fishing gear, the latter to ensure it is not used elsewhere. The determination of the disposal of the vessel in the case of conviction varies considerably between countries. The level of fines is the key point of interest. There is a tendency for fisheries laws to discriminate between national and international fishers. This can, in itself, create difficulties in the courts. It is sometimes best, and often recommended, that financial penalties for offenses be standard for the offense despite the offender. The level of deterrence can be determined by the judiciary in noting the economic impact of the level of the penalty on the violator. This implies that fines would be standard, but they would be worded in a manner whereby the judiciary would have the flexibility of determining the penalty within a large enough bracket which could serve as a deterrent to all fishers using the zone.

The education of the judiciary to meet the above noted objectives can often be influenced through a request for assistance from the judiciary at training sessions for fisheries personnel, and round table discussions of the effectiveness of annual operations, as well as seeking advice with respect to the writing and application of the fisheries law. It must be recognized that fishers, will ignore what they believe to be unnecessary legislation. The support of the judiciary in these cases can be noted by the size of the fines levied for these offenses.

National police, and possibly other agencies, such as natural resources, etc., often have an important role to play in fisheries MCS. In several countries, these agencies are the implementing arm of the MCS strategy. The cooperation of these agencies and their personnel, whether directly involved in MCS for fisheries or not, is important for cost effective intra-government operations. It has been seen in the past that, in several cases, individuals who habitually violate the law are often multi-purpose offenders in that they are sought by more than one agency. Inter-agency co-operation is most cost effective in these cases. Also for consideration in developing countries is the potential benefit which can accrue from providing the powers of fisheries enforcement to other enforcement agencies. This then expands the potential surveillance force of the country. The contact these agencies have at the community level, if utilized appropriately, can assist in establishing the linkages required for successful operations.

There is however, an obligation which follows this delegation of powers which is often forgotten. It is the obligation to the fishers that the individuals surveying them and enforcing the laws have the appropriate training and knowledge of these laws. It can prove detrimental to the credibility of fisheries if enforcement personnel do not have the training and knowledge of fisheries laws and law enforcement procedures. If the training is not possible, it is best not to delegate the powers for fisheries enforcement too broadly.

The port authorities have a considerable impact on the fishing operations of both national and international fleets. The actual, and potential, control and benefits this agency could provide for fisheries and vice versa are significant. Recognizing that fish is a perishable product and the needs of fishing vessels are somewhat urgent, the liaison between agencies can facilitate the exercise of the port requirements and the fisheries inspections and briefings in a timely manner to expedite the process for the fishers. This would minimize their concerns regarding port visits and potentially enhance port and fisheries capability and possibly the economy of the port. Port authority powers over vessels can also assist in the surveillance component of fisheries management if this is required for inspections or other enforcement procedures, including detention of the vessel. It is suggested that, although port authorities have other priorities, good liaison could result in support for fisheries operations.

4.5 **Consultation and liaison**

Success of fisheries MCS operations in almost all cases has been due to the liaison between parties with a vested interest in the fishing industry. Clear delineation and acceptance of roles, responsibilities and obligations has made the process easier for all concerned. The harvesting of any natural resource by strong, independent business persons has always been a challenge and nowhere is this more evident than in fisheries. Successful MCS operations have reflected the absolute requirement for all participants and individuals with input into the process to understand and accept the management plans.

4.5.1 Fishing industry inputs

The purpose of rational, sustainable fisheries management, and subsequent MCS activities in support of this concept, is for the sole purpose of providing continuing benefits to the fishers of current and future generations. It would seem apparent then, that key to any discussions regarding the fishing industry would be the fishers themselves. Unfortunately, there is often little input from these, the grassroots clients and beneficiaries of the fishery. It is no surprise that, after many years of neglect and little input into fisheries management fishers are suspicious of government officials. Fishers prosecute the resource for financial gain and they do this armed with the combined knowledge of generations of practical knowledge passed down through families and the folklore in the communities. There is a need for respect and partnership of government officials and the fishers themselves in the development and implementation of fisheries management plans in the future.

This linkage with the fishers can often be assisted through liaison with fisher groups, unions, or cooperatives, where these are in existence. Where these are not in place the community organizations themselves can serve as a good link to the fishers and their families. When the fishers become confident that the intent of the government is really in their best interests, the government will have a very influential and powerful ally in implementing its plans.

Fishers who are organized appear to be more successful in ensuring their input into fisheries management plans. Most governments have extension officers of some departments to liaise with communities. These individuals can greatly assist in this process of establishing links with the communities for fisheries matters. Trust and cooperation are key to this process. Government initiatives to assist fishers in communities to organize themselves for fish trade purposes may also be viewed positively and contribute to this overall exercise of fisheries management.

The input of bigger fishing enterprises with contacts on the international market for fishing vessels, joint agreements and trade are also very important in this exercise. There is a caution from years of practical experience, however, in placing too great an importance on one sector of the fishery, while neglecting the impact other sectors have on the industry (usually at the expense of the artisanal, domestic fishers). The larger fishing enterprises should be encouraged to provide their input, and in fact will probably do so without urging due to their investment in the industry. The challenge will be determining the balance between interests of large fishing industry interests and the more scattered local artisanal fishers.

4.5.2 Inter-ministerial liaison

The large number of ministries interested in the ocean sector and particularly fisheries creates the potential of administrative complexity. A listing of these ministries may assist in focusing on their respective roles and input that the Fisheries Administrator can expect from government partners. These include the Ministry of Justice with its judiciary, municipal and federal police agencies, port authorities, national defense, customs, immigration, health, foreign affairs and the fisheries department itself.

If one breaks down the interests, it can assist the Fisheries Administrator in the determination of the points of contact and timing for each party in the MCS process. The *Ministry of Justice and the judiciary* need to understand the fisheries management objectives, policies and importance of the resources and habitat to the State. Second, they need to have a good understanding of the intent behind the fisheries management plans and the MCS procedures to reflect these in the fisheries legislation. It is worth fostering the relationship between fisheries and the judiciary to obtain their assistance with fisheries legislation and its implementation.

The cooperation of the *federal and municipal police agencies* is an essential component of MCS. These agencies are often the operating arm of the surveillance component. Whether or not this is so, the cooperation of all enforcement agencies through shared databases and shared resources has proven very cost effective in the past. Where priorities can be arranged to coincide, it is always effective to join forces for enforcement purposes. Another consideration for fisheries with respect to other enforcement agencies is a delegation of powers and authorities for enforcement to the field personnel of these agencies. There is an obligation to the fishers resulting from this possible delegation of powers, however, and that is the assurance of appropriate training in fisheries enforcement, laws and procedures for these enforcement officials. The federal and municipal police agencies have contacts which can also be very useful in establishing the appropriate links with fishers to involve them in fisheries management and implementation planning exercises.

The *Port Authorities* can be of considerable assistance in fisheries MCS, both for monitoring and surveillance activities. Port State control is becoming a more popular initiative and a multidisciplinary port authority can facilitate coastal State port inspections, briefings, and transhipment of fish with a minimum of bureaucracy and maximum of control at low costs. The Port Authority can also assist in obtaining information regarding the vessel and its operations and can benefit from the sharing of the fisheries database with respect to vessel movement in the zone. Sharing of resources for monitoring and enforcement purposes could also be considered here as being potentially cost effective.

The *Ministry of National Defense* is very interested in fisheries operations under its sovereignty and State security mandates. There have been cases where military resources, having been seconded to fisheries and under fisheries control, have proven effective for MCS activities. The United Kingdom fisheries services operated in this fashion for several years in England and Wales. In Scotland an executive agency under the Department of Agriculture and Fisheries in Scotland (DAFS) operates the government owned patrol and research fleet. This latter option has been found to be cost effective.

It must be stressed however, that MCS for fisheries is a civil police action, consequently, the use of the military as the executive power of the State is not really appropriate for direct involvement in this task.

The military can however, in the course of their regular duties, assist in fisheries matters. The national defense departments of Australia and New Zealand have proven to be very effective in their air surveillance role for FFA over the years and have become an integral component of their system. On a smaller scale, sub-regional or regional organizations may be able to secure resources for similar services, or purchase their own aircraft and use seconded military personnel for flights and maintenance. The sea component of the military has not yet been found to operate efficiently or cost effectively in a support role for fisheries in most MCS situations, due to the cost and bureaucracy of these heavily manned military resources. These high level military resources have been used by states in the apprehension of foreign vessels, but it has been expensive and control of the operation has been lost to fisheries personnel in this "civil police action". Only secondary tasking of resources from the military should be contemplated for fisheries MCS activities. The use of the military sea component is ineffective and inappropriate for MCS activities respecting domestic vessels and is difficult for foreign vessels. The sharing of the fisheries database with the military might also assist them in their priority sovereignty mandate.

Customs and Immigration are being considered together due to their similar interests of control of goods and persons into, and out of, the State. Again cooperation for surveillance services and sharing of resources can be cost effective, but only when priorities coincide. The American situation, where the U.S. Coast Guard enforces customs, immigration and fisheries legislation, creates difficulties, as the drug situation in the U.S. demands the resources on a priority basis. Fisheries thus have a low enforcement profile during these periods of drug interdiction. This drug interdiction requirement also influences the design of the vessels, which are usually inappropriate for fisheries where sea-keeping is a priority. One advantage of cooperation with these agencies is the sharing of surveillance information and secondary tasking for possible back-up support for surveillance activities.
The sea aspects of these two departments are not complementary to fisheries and would most likely place fisheries personnel in a more hostile environment than would be appropriate for their normal function. Both of these agencies will also have an interest in fishing vessels during their port visits. Coordination with fisheries could facilitate these operations for the benefit of all parties.

The *Ministry of Health* usually has an interest in fishing vessels and the fishery for two main reasons; first, for international fishers, the state of health of their vessels and crew when they enter port and, second, for the standard of the product that is landed for domestic consumption and possible export. The Fisheries Administrator can gain assistance from these officials regarding fish product inspection concerns by cooperating on these matters.

Foreign Affairs has a considerable interest in fisheries, especially where it involves international fishing partners and negotiations. It is usually foreign affairs which takes a lead role in these negotiations and the challenge for Fisheries Administrators, as noted earlier, is to ensure that the bottom line for fisheries management and MCS control is not compromised in negotiations. Key to the negotiations is the principle of lowest cost, effective MCS to conserve the stocks. Any negotiated change that increases the costs of MCS should result only if there is a corresponding increase in benefits, preferably financial, from the international party to offset the increased costs for implementing MCS. An example is the case where the international fishing partner wants to have only one vessel with the fleet commander come to port to pick up the licenses for the full fleet. The coastal State thereby loses the opportunity to verify the fish onboard each vessel for later calculation of the total fish caught in the zone. The offsetting agreement could be that the fleet commander's vessel would carry observers, paid by the international fleet, to the other vessels, where these representatives would estimate fish onboard prior to the vessels commencing fishing in the zone. An alternative would be an agreement for a fisheries officer to be transported by the fleet vessel to each other vessel, and then returned to port. A final alternative in this case could be an agreement for a grant of the equivalent funding to permit the patrol vessel to deliver the licenses to the vessels. In all cases, the vessels should not fish until the fish onboard have been verified by a fisheries official, or representative. These alternatives preserve the principle and importance of the first verification of fish onboard the vessel prior to its operations in the country's zone.

It is important that each of the MCS activities is clearly understood by the foreign affairs negotiator. There may be a tendency on the part of the international partner to attempt to negotiate other options which decrease the effectiveness of fisheries MCS.

The *Fisheries Units* themselves, especially in the remoter areas of the country, need to be kept abreast of the fisheries MCS strategy. The heart of successful MCS operations in the field is the communications system. A telecommunications network is almost essential for MCS operations in the current fishing environment. Lack of information creates insecurity, concern and difficulties in supporting government policies and explaining these to the fishers. It is necessary for a credible relationship with the fishers that all the field units are fully aware of the policies and the rationale behind these decisions. On the other hand, it is also important that fisheries officials in headquarters also realize the benefits of the information which can be provided by their field units for all aspects of fisheries management and MCS operations.

It was stated at the beginning of the paper that fisheries management and successful MCS operations include gathering and analysis of biological, social and economic data, decisions regarding the overall fisheries management strategy, then the legislation to form the base for implementation of this plan and, finally, the MCS operations. This requires multidisciplinary management and input from several sources in a fisheries department. This is one of the greatest internal challenges that Fisheries Administrators will face. Each section and each individual has an important part to play in the fisheries management exercise. The team-building effort required to gel these diverse components into one team for the MCS exercise is considerable.

4.6 MCS plans

MCS plays key roles in monitoring and data collection, forming the legislative base for successful implementation and is the action arm of the policy. This makes it apparent that MCS must be an integral part of planning from the very commencement of fisheries management planning. Once it is determined which control mechanisms (effort, catch, quota, area, fish and mesh sizes, or a combination of these) will be utilized for fisheries management, then the MCS officials can assist in planning an enforceable fisheries management plan for both the domestic and international fleets. MCS plans should be developed consecutively for each fishery, or group of fisheries. This will permit analysis of the legislation required and its enforceability. The assessment of MCS resources to be deployed can also be determined and all parties then can realize the pressures being placed on this infrastructure. The final package will, when assessed, show the actual potential of meeting these requirements.

The international fishing plan, if international fishing is to be permitted, can follow a similar exercise, with one major additional input, the results of negotiations. Many factors for the negotiations have already been noted, the most important being the maintenance of control of the fishery in a non-discriminatory manner for both fishing fleets, national and international. The rule of keeping MCS simple and exacting benefits and remuneration for any divergence from this principle to recover the additional costs to the State is one to keep in the fore during negotiations. It is strongly recommended that senior MCS personnel be included in the fisheries negotiations to assist the chief negotiator in the technical aspects of the MCS portion of the agreement. This can prevent difficulties in enforcement and misunderstandings in the implementation of the agreement. The impact of the negotiations then needs to be reflected in the fisheries legislation and in the MCS plan. The analysis and determination of the resource requirements and strategies to implement the agreement with an acceptable assurance of compliance is the following step in this process. Senior fisheries managers should then look at the total picture for MCS requirements for both the domestic and international fisheries and establish priorities on the use of the resources to permit MCS annual planning. In this manner, all senior fisheries decision makers, up to and including the Minister, can be briefed by the Fisheries Administrator as to the expected deployment of MCS resources and the assessment of risk and consequences of the implementation plan. It must be accepted that there will not be enough MCS resources to fully address all aspects of the implementation of the fisheries management plans.

5.   MCS OPERATIONAL PROCEDURES

It is intended in this section to look at some actual processes and core points for coverage in MCS operations, which have been used with some success in various parts of the world. The section will touch on data collection and components for the system, boarding procedures, catch verification, navigational positioning, inspection procedures in port and at sea, transhipment verification, planning patrols, evidence gathering and handling, and prosecution procedures.

5.1   Data Collection

In the case of **data collection**, a major component of the **monitoring** aspect of MCS, there are two preliminary and one follow-up activity in which fisheries officials become involved. They are the fishers' licensing information, the vessel registration system, and the catch or effort monitoring system. The fishers and vessel monitoring databases, whether manual or computerized, will necessitate an initial census of fishers in the country and specific information on international fishers. With the initial domestic census, the following information is usually included as it is useful for fisheries and MCS planning. The fisher information includes the name of the fisher, home address, age, experience in years and type of fisheries, whether fully or partially dependent upon the fishing industry, position in the fishing industry as a vessel owner, operator or crew member and a general average income from fishing. Further monitoring information by fishing trip, if included, is intended to collect and cross check fish catches, broken down by species and weights, area of fishing activities, time of fishing and finally, the returns from the fishing activities which can be used to calculate the efficiency of the fishing unit.

Vessel registration is intended to collect data which can be cross-linked to the licensing system, such as the description and size of the vessel, home port, call sign, where fish are landed, catch capacity in terms of hold and fishing gear type and capability, experience and efficiency as a fishing unit, the age of the vessel, outfit including communications, navigation and fishing gear, and processing capabilities, if any. The potential of regional and other international cooperation on developing and implementing standards for vessel registration have already been noted.

This census of fishers and vessels active in the domestic fishing fleet needs periodic updating, a task that can be facilitated through annual licensing procedures and appropriate report procedures during the fishing season.

Most of this information is used to ensure that the fishers and the fishing units act in conformity with the agreed fishing plan, but it is also used by the fisheries biologists to assist in stock assessment exercises. Fisheries economists and sociologists can use the information to determine the importance of the fishery to the national and community economy. The sociological profile of the fishers and their communities can also assist with enhancement of their position in the social and economic scale and provide support facilities where needed. Infrastructure, communications and data networks, is needed to support these data collection activities.

5.2   Fisheries Patrols

Fisheries patrols can be made more cost effective if planned with the view of integrating the surveillance resources to achieve the best results. All patrols should commence with pre-planning, a briefing of key participants to ensure that there are no surprises, the actual patrol, and a de-briefing on completion, with appropriate documentation for record purposes, or follow-up action as required.

Land patrols up rivers and along lakes and the coast can be effective if focused on fishing activity, areas of illegal activity or zones where the fisheries resources are particularly vulnerable to over exploitation by licensed and non-licensed fishers. Coastal areas, where domestic fishers operate and can be seen from land, can be watched for incursions by larger vessels not authorized to be in the area. This information can be relayed to coastal sea resources for action as appropriate.

Air patrols are most effective if the fishing areas denoted by season and species are known in advance. The air surveillance capability must be taken into consideration to determine the areas of priority, as endurance time of the aircraft will determine how many priorities can be addressed in a single patrol. The aircraft crew should be briefed on the patrol area and the expected activity in the zone, as well as the priority activities for surveillance. The air crew should also be provided with a summary of the vessels, their markings and authorized activities which can be expected in the patrol zone. Examples of patrol priorities might include a closed spawning area as the highest priority, then the aircraft might proceed at altitude to save fuel direct to that area and commence its lower level patrol activities from that point. This forward planning reduces fuel consumption and increases patrol time in the desired zone. Random patrols, without a focus, have been found to be less cost effective than directed patrols for a specific purpose. Another priority might be an area of fishing concentration where local fishers have noted incursions of offshore vessels into their zones at night with resultant gear destruction. Areas of heavy fishing concentrations where non-licensed vessels may hide during fishing operations could also be a priority. If stocks migrate close to the edge of, or beyond the fisheries waters of the State, there can be a temptation of offshore vessels to follow the fish into the zone if they believe the risk of the activity is small and they will not be apprehended. Air surveillance provides the front line information for the deployment of other more expensive resources, such as offshore patrol vessels.

Coastal patrols at sea are most effective if smaller patrol vessels can be pre-deployed to areas of fishing concentrations and operate from a base in this area to provide a timely response to fisheries conservation needs. In this manner, the patrol vessels can shadow the coastal fishing fleet for data gathering and verification and surveillance. The presence of a fisheries patrol vessel can also contribute to fisher safety, but this can be abused. There have been cases where fishers took turns to raise safety concerns to get a tow from the patrol vessel, thus putting the latter out of the patrol zone for a period while all the remaining fishers prosecuted the fishery in a spawning area.

Offshore patrol vessels, if it is decided that these are to be utilized, are best deployed to areas of concentrations of offshore fishing. Air surveillance is the primary tool which can detect area violations and unlicensed fishing activity. The patrol vessel can then be called to the scene, if necessary. More cost effective is the possibility of using diplomatic channels to bring the vessel to port, but this may not always be possible without greater international pressure than that which a single state can exercise.

It is obvious that each patrol vessel should have the necessary accoutrements to carry out assigned duties, e.g., copies of regulations and communication equipment.

5.3   Boardings

The decision as to whether it is safe to board due to weather is that of the master of the patrol vessel. The fisheries officer is the leader of the boarding team and as such it is the officer's decision whether the boarding party will actually board a particular vessel. On fisheries patrols there should be no doubt that the vessel is for support of the fisheries activity and hence the fisheries officer is in operational command of the patrol.

The fisheries officer must ensure that the master and boarding team are briefed on the boarding procedures and expected support and communications for the boarding. The fisheries officer should ensure that the team is aware of the latest data on the vessel to be boarded, its license and fishing capability, design of the vessel and, if known, the route to the bridge of the vessel. The fisheries officer should have the latest pre-patrol data on the catches and last reports from the vessel for verification with onboard records. The fisheries officer than checks that the boarding team is properly equipped with documents, inspection equipment, such as gear measuring devices, safety and communications equipment.

The fisheries officer should ensure that notes of observations of all activities on the vessel to be boarded and responses to communications are made from the moment the vessel is sighted. These observations may prove very useful if there is an alleged violation and can assist the prosecutor in developing the case. The notebooks of all involved are therefore useful tools for recording events and observations and should be used accordingly. Dates, times and events should be recorded faithfully by all fisheries officers and patrol personnel on the vessel and boarding team.

Many fisheries officers make up their own package, in addition to that issued, to assist in the facilitation of the inspection. Some states have formalized these into quick reference field manuals for the officers. They may contain information on the fisheries regulations, commercial fish identification, fishing gear, common phrases translated for use in questioning the vessel master, special vessel identification markers for each country active in the zone, check lists, communications information and signals for use in special situations, measurement conversion graphs, and others. The fisheries officer should be aware if other fisheries department officials are aboard and what their assigned tasks are.

The boarding brief should delegate the activities of each member of the team for the duration of the boarding and any special instructions in the case of hostilities or resistance. If hostilities break out during the boarding, each team member should be aware of the disembarking procedures. The communications link to the patrol vessel is essential if there is potential for a less than friendly reception of the boarding crew. There are cases when the boarding team consists of the fisheries officer and an assistant, but these are rare, a minimum of four persons should be in the boarding team. In cases when the reception may not be friendly, a full (minimum of six persons) and equipped boarding crew is advised. This would be a team of approximately six persons; the fisheries officer, a navigating officer and an engineer with two or three crew, if these numbers are available. There is protection in numbers on occasion.

The patrol vessel and boarding boat should be identified as being on fisheries patrol in accordance with the Convention on the Law of the Sea. The procedures for halting the vessel and responsibilities of the fishing vessel master to receive the boarding party should be well defined in the fisheries legislation of which, if the vessel is licensed, the master should be aware. There are clear procedures in the international code of signals for visual and radio communications to the vessel to stand by to be boarded. The alpha, numeric flag signal, Sierra-Quebec-Three (SQ3) is the international signal indicating "Stop or heave to: I am going to board you" and may be used to signal the intent of the boarding team to the fishing vessel. The master of the fishing vessel may communicate that he will complete hauling the net or other fishing operations prior to boarding. If boarding would result in potential loss of fish or gear, the fisheries officer should respect this request. The master of the fishing vessel should steer a course to provide a lee side for the protection of the boarding party while climbing the boarding ladder. The fishing vessel may stop, but experience has demonstrated that it is actually easier for the boarding boat and team to get on the vessel if it keeps some way on the ship in the range of a few knots. It is possible to board comfortably at fishing speeds, but speeds above 10 knots become more difficult, even for an experienced team.

Once on board the fishing vessel, the boarding team should move quickly to the bridge area. If a less than positive reception has been encountered, one crew member should remain at the boarding ladder. On the bridge, the fisheries officer should ask for the captain of the vessel and provide government identification which clearly shows that he is an authorized government fisheries official. He should request the assistance of the vessel captain for the inspection of the vessel. At this point there should be an indication if the boarding will encounter difficulties in cooperation or not. If no difficulties are indicated, the fisheries officer should carry on with the inspection.

If there is an indication of difficulties, a potential hostile boarding, then the team should fall into the operational procedure which should have been covered during the pre-boarding brief as to the action to be taken. These procedures will undoubtedly vary considerably, depending on the equipment and training of the boarding party. They could range from departure and diplomatic negotiations to securing the bridge and the vessel and sailing it to port. It is naturally best if the vessel crew can be convinced to cooperate. If threatened, the boarding crew must have the authority to take whatever action is deemed necessary to protect themselves. The conclusion of the boarding is a reverse of the procedure for getting aboard, with the fisheries officer usually being the last of the boarding party to leave the vessel.

Linguistic differences tend to pose an initial concern for both fisheries officials and the vessel master. There are two common solutions to this problem, one is to require the use of reports, logbooks and such that are issued by the State, in the language of the State. The responsibility then rests with the international partner to translate the information. Some countries and regional organizations complement this first step, or use a second step separately; that is, to obtain copies of the relevant reports, logs and documents from the international partner and translate them for the State's officers. A further tool is a small handbook of questions in the various languages with common numbering system so that the appropriate numbered question can be asked. This procedure has been used effectively by many international fisheries organizations. Most international organizations could assist in developing such materials. It is advantageous if there is a member of the boarding team who has the capability of understanding the language of the vessel.

5.4 Inspection Procedures

This section considers the actual *at-sea inspection* of the vessel. Following the preliminaries of introduction and identification of the fisheries boarding official to the master of the vessel, it is common to request the license and all the fishing logs related to fishing, transhipment, processing and storage of the fish. The ship's navigation log is also used as a check of position information. The engineering log is usually secured at this time as well to check the use of mechanical machinery. If there are any other fisheries department representatives or officials onboard, it is recommended that they be sought for introduction. If it is possible, a quiet and private discussion with these persons prior the inspection is recommended to ascertain if there are any fisheries matters that will warrant more careful scrutiny. An inspection focuses on data gathering for two purposes. The first is for surveillance of the fishing operations to determine compliance with the terms of the license and legislation, and the second is to gather data for the monitoring aspect of MCS and fisheries management. The verification of the logbooks should be sufficient to reconstruct the fishing activities of the vessel since entry into the jurisdiction of the State.

Most countries design their own boarding format to meet their data requirements and to facilitate computer entry, if available, and data cross-checking with other reports. Almost all reports have a section to identify the vessel, its license, confirm the name of the master, verify its activities over the period from entry into the zone, or from the last inspection. In the case of the latter, the report of the inspection should also be present. The boarding report normally has a section to record a summary of catches by species, effort and areas fished. A section on production, storage and transhipment is also usually included for recording, if appropriate.

Fisheries officers usually commence their inspections with a check of the licensed activity and verify that the positions and activities in the navigation/ship's logbook confirm that the path of the vessel is in compliance with the fishing license. Any reports made from the vessel to the fisheries department are noted and checked against their source. It is sometimes advantageous if the fishing log and catches, when summarized by the fisheries officer for the inspection form, are broken down into the same periods as that required by the State for the vessel reporting to the department. This facilitates cross-checking on arrival in port. The transhipment of fish by species and date is noted as well as the name of the vessel receiving the fish. Armed with this paperwork, the fisheries officer then attempts to verify the figures presented in the logs of fishing, production and transhipment through an inspection of the processing plant and the storage facilities. There are various conversion factors used in determining the production and storage of each type of fish and fisheries officers will, over a period of time become very proficient in determining these. In the initial stages of inspection of vessels, the factors used by the vessel master may be those used for the conversion of the processed fish back to the whole round weight of the fish. The inspection report usually includes space for the vessel master to comment on the report and then both the master and fisheries officer sign the report, the latter leaving a copy for the vessel.

The difference of at-sea inspections compared to *port inspections* has been noted. The latter does not permit the ability to see the navigation, fishing and processing equipment in action. The advantage of the latter is the ability to check the fish onboard the vessel in a safe and stable environment with a higher potential for accuracy.

5.5   Verification of Catches

One of the challenges of inspections when the management strategy is based on catches and quotas is the *verification of catches*. There are many different aspects to this process, the first being the type and processing, or product form of the fish in its final storage. It may be seen as a simple matter of counting boxes, but in a large fish hold with a capacity of 500-700 metric tons of product from several species, the task becomes onerous. In most cases, verification of catches becomes a mathematical problem.

The independent and accurate estimation of total catch by species on a set-by-set basis is the fisheries officer's/observer's most basic and important function. Often information from these estimates of catch composition provide the only reliable estimate of removals in certain fisheries, as traditional recording methods, such as logbooks, have generally yielded incomplete data.

Catch estimates must be both independent of those derived by the captain of the vessel and as representative as possible of what is occurring in the specific fishery. Domestic vessels, the catch of which is off loaded in a domestic port, are much easier to verify than a large offshore international trawler. Measuring techniques appropriate to each country can be assessed using various sampling methods on domestic vessels and verifying them against weigh outs on landing. The following provides a few general methods in use today which may be of benefit and can be re-configured for local use.

Estimation of total catch of trawlers

While a direct weighout is the best verification of the amounts caught, it proves impossible for most of fisheries due to the large catches involved. A number of estimating procedures have been developed to verify the total catch. The two basic methods commonly used on trawlers are:

1. Observation of the catch in the codend
2. Volumetric calculation of fish pre-processing holding bin capacity.

There are two additional methods to both provide the estimates of the total weight and to verify the previously made estimates of total catch:

3. Volumetric calculation of fish holds and use of production figures.

Further explanation of these methods follows:

1. <u>Observation of the catch in the codend:</u> An estimate of the total catch may be obtained by knowing the capacity of the codend and approximating the percentage of it filled with fish, by taking a volumetric measure of the catch in the codend or by breaking the codend into smaller volumes to estimate these sections. The rough and ready estimate comes from a visual calculation. Vessel masters usually place the lateral strengthening ropes on their trawls at stress points along the cod end. These relate to a very rough approximation of one and one half to two tons of fish per strap when the cod end is full. This gives the fisheries officer a visual estimate before asking the vessel master for an estimate. It should be

remembered that the master has been estimating this way for several years and consequently, the estimate received from the fishing master is probably much more accurate. This visual method is the roughest and most inaccurate for estimating catches. The sampling of smaller sections of the codend usually yields the best results.

A basket, of known volume and weight, should be used to take samples of the fish in the codend. The vertical strengthening straps divide the codend into a number of sections. Each section should be broken down to an estimate of the number of sampling baskets (of known volume/weight) of fish contained therein. Due to the tendency for fish to pack more densely in the aft of the codend, each section of a codend should be treated separately. For smaller catches, the number of sampling baskets for the total volume of the catch in the codend should be estimated. Thus, the estimated number of baskets should be multiplied by the average weight of fish per basket, allowing for the variations in catch densities.

2. Volumetric calculation of pre-processing holding bin capacity: Catch is usually stored in holding bins prior to processing. This presents a perfect opportunity to verify the initial estimate obtained by viewing the codend.

The volume of the holding bin has to be determined and multiplied by the density of fish to calculate the capacity (the density of fish can be easily calculated using a small container/sampling basket). Once the holding bin capacity is known, the amount of fish in the bin can be determined by estimating the percentage of the holding bin filled with catch.

3. Capacity of fish storage hold and use of production figures: The capacity of the storage area can be used to verify the initial estimates. The total fish hold capacity can be obtained by interviewing the captain or from ship's drawings and previous inspections.

Estimating techniques vary considerably, depending on whether the fish storage is wet or dry. If it is dry and the fisheries officer can get into the fish hold, it is a matter of sampling the fish, probably frozen, in the boxes for product form and average weights of the product in the box. This should be done several times for each species and the box weighed separately for later subtraction of its weight. If possible, the number of boxes should be counted and the weights calculated using the average for each species. If it is not possible to physically count all the boxes, the vessel drawings should provide enough information for the officer to estimate the number of boxes in the fish hold using a mathematical formula to compensate for the vessel contours.

This estimate can be cross-checked against the number of boxes the storage manifest states there should be in the fish hold. The actual content by species will be very difficult to obtain without a physical check; consequently the storage manifest, or log, may be the only documentation available to address this point. It should be remembered that the purpose of a random sample and catch estimate is to assure the officer that the documents are accurate. An estimate will undoubtedly result in a difference from the records and the actual numbers of boxes in the fish hold, but it should be similar to the records. An officer's judgement is called upon before making a decision to bring the vessel to port for off loading due to the expense and consequences of such an action. If the latter is not accurate, the captain's estimate, coupled with a cross check of the catch and production logs should assist in making this decision and determining the potential level of inaccuracy in the records.

If the species has been wrongly noted in the log, it will be difficult to determine, but if the officer checks a random sample of the labels on the boxes, and where possible the product therein, it may be possible to ascertain if the practice of misreporting on the quota of a higher priced species is common. If misreporting of catch is found, it is grounds for the vessel to be brought to port for further investigation.

Assuming, however, that fish hold measurements and access to the hold are possible, the production log and storage logs become further checks as to the estimate in the hold. On factory-type vessels, the units of production (i.e. boxes) from a specific set can be tabulated, multiplied by the unit net weight and converted to round weight to check the accuracy of the initial estimates. The production log should note the fish processed to the current date and the storage log should note the fish boxed and stored in each fish hold. It becomes a matter of calculating the daily totals for the period, cross-checking these against each other and the estimates in the fish hold to verify if the records seem to be reasonably correct. It should be noted that the processed weight/production figures, using this method, are being used to verify the initial estimates of the retained portion of the catch. It does not include the weight of fish and offal discarded.

Catch estimation for other fisheries and product types

Other fisheries, using different gears, may necessitate a totally different approach to catch estimation. Below is a brief description of procedures used to estimate the catch in longline and purse seine fisheries.

1. Wet or salt fish: The estimates for wet or dried fish such as salted fish, are very difficult to obtain due to variances in the types of salted fish; light salted and heavy salted. The duration of the catch in the salt, density, and also fish hold capacities again come into play in these estimates. Some countries have attempted different methods for these estimates using volumetric methods, salt densities and hold conversion factors. The Canadian fisheries officials at one time developed a computer programme for estimating salt fish in the hold of a vessel, but the luxury of computer access is not always available. This does however, show the difficulties encountered in making catch estimates.

2. Longline fishery: The nature of this operation does not allow one to see the entire catch at once. Fish are coming individually on board and the number of individuals may be easily counted and multiplied by the average weight of fish (determined through sampling) to obtain the estimation of the total catch of the species. Occasionally, the fish is stored in a bin, or pen, on deck before processing. This gives an opportunity for volumetric calculation.

Some longline fisheries (e.g., tuna, shark) present an opportunity to weigh all fish caught, provided appropriate scales are on board. In the absence of those, other vital measurements such as fish lengths can be obtained and translated into corresponding weights using tables.

The estimate of the weight of large fish which are frozen whole, such as tuna, is very difficult to obtain due to the variance in the size of the fish and the storage. A sampling of the fish can produce an estimate for extrapolation, but these estimates are very rough. The best figure one can expect is from the receipt of weigh outs at the port of landing and through a cross check of the fishing and storage logs.

3. Purse seining: In the case of wet fish storage, such as refrigerated circulating sea water (CSW) systems for herring as an example, several methods have been used. Herring, and possibly other fish, have a tendency to move to the bottom of the tank in a CSW system. One method is to drain the tank down to the top level of herring and dip the herring to see how much is left. This can be time consuming and the water has to be replaced to assist in off loading, as the herring are pumped ashore to prevent damage to the fish.

A second method is to weigh the herring as it is removed from the vessel after draining, but this is not always possible if the inspection is at sea, or in a port other than that designated for off loading.

A third method designed in cooperation with herring seiner captains themselves is most common. This method requires a pre-season calibration of the fish hold with the vessel in a stable, upright position. Marks are then placed on the bulkheads of the fish hold to indicate the level of water and fish in the space. These marks then are equated, through pre-set volumetric calculations using a common fish density, to a calibration card that provides an estimate of the amount of fish in the fish hold. This method requires the fisheries officer to check the for-and-aft trim of the vessel as well as the list. The vessel master then attempts to bring the vessel as close as possible to an upright position, possibly by swinging a boom or pumping and flooding tanks. The officer then takes a weighted tape attached to a large screen apparatus and lowers this into the fish holds until the screen rests under the water on top of the herring. The readings on the tape and the fish hold calibration table are then used to provide an estimate. This is done at least four times in each fish hold and then averaged. This dipping process is simpler than other methods and has received support from the industry as it is timely and easily carried out without considerable input from the master or crew.

Estimation of catch composition

While estimating total catch is not an easy task, an accurate determination of catch composition may present an even bigger challenge. Four basic methods have been developed to derive a breakdown of catch by species:

1. Actual weighing of the catch by species: This method can be utilized for small total catches or with small amounts of bycatch species present in the total catch.

2. Extrapolation from the surface area occupied: This approach involves the estimation of the percentage of the known area (usually surface of fish as it rests in the bunker) occupied by each species in the catch. These percentages are applied to the total estimated catch to obtain the weight of individual species. This method should be used with caution, since some species may not appear on the surface, due to different densities.

3. Extrapolation from the random sample: A random sample is gathered from the catch and the estimate of the percentage weight of each species is made. This percentage breakdown is then extrapolated for the entire catch. This method has proven to be very effective for catches composed of fish of uniform size.

4. Monitoring the catch exiting the fish holding bin: This approach involves tallying an estimated weight of bycatch species exiting the fish holding bin. The figures are subtracted from the total estimated catch to arrive at the estimated catch for the major species.

Estimation by production category

Once an estimation of the total catch and its composition by species is made, the final estimation, by production category, is performed. It involves the determination of a round weight of species **retained** for further processing and round weight of species **discarded**. Discards should always be estimated, or if possible weighed. They have to be subtracted from the total species weight to determine the retained weight.

A final variable in the calculation for fisheries records is the conversion of the product form back to the round weight of the catch, for this is the figure to be used for determination of the total catch which has been retained. The conversion factor from whole fish to product form depends on the efficiency of the processing equipment. Sharp and well maintained equipment, or experienced manual plant workers, can make a considerable difference in the conversion factor of the final product form.

The maintenance of the processing machinery can increase production by a significant percentage, and thus the conversion factor from processed to round weight will reflect a considerable difference in the estimate of fish onboard the vessel. If the factor for fillets is estimated to be 1.4, then 20 tonnes of product would convert to 28 tonnes of round fish. If however, the real efficiency of the plant is 1.6 for fillets, this same 20 tonnes of product becomes 32 tonnes of round weight, a significant difference. It is necessary to carry out such calculations to determine the catch efficiency of the vessel and to calculate the catch and effort and the portion of the quota caught by the vessel.

There are several variables in rough estimating methods which must be noted. Some of these include the space occupied by the fittings in the hold, the space between boxes, the contour of the vessel bottom and estimates of capacity for storage of boxes, especially if they vary in size and weight. There is also the variance in the weights of the boxes themselves. The difficulty in determining the type of fish once processed, such as fillets, is another potential problem. These points are noted to emphasize that without an off loading, it will be very difficult to make an accurate estimate. It is for this reason that the fisheries officer must use judgement when making the estimates, anticipate these variables, and consider the final figures carefully before making a decision regarding bringing the vessel to port under an allegation of misreporting. The latter creates a considerable loss of fishing time and cost to the large vessel fisher, and without justification, can impact severely on the credibility of the Department to carry out its duties effectively. The bottom line is *tolerance*. This may vary according to the circumstances at the time, the amount of fish onboard, the value of the fishery, the location of the vessel, the past performance of the vessel and master, etc.

5.6   Verification of Position

The *verification of position* of the vessel is a key requirement for any fisheries sighting, inspection, or prosecution. Surveillance aircraft are usually equipped with highly accurate positioning systems, many linked to automatic photographic equipment that shows the position of the aircraft or target at the time of the picture. This has been accepted in courts in many countries as evidence of the position and activities of the vessel, provided the continuity of the evidence has been preserved. In the case of at-sea inspections, it is common procedure to take a position on the boarding vessel just before conducting the boarding. The fisheries officer should ask the vessel master for the position of the vessel upon boarding and if there is a variance, the officer should watch the master take another fix. If the officer is not satisfied, and does not hold appropriate navigation papers, it is recommended that an appropriately certified deck officer from the patrol vessel take a fix on the boarded vessel. One deviation that some masters have used to avoid charges is the removal of a fuse from the equipment, claiming that the equipment is faulty. A trained navigating officer can usually determine if the master is attempting to avoid responsibility for possible illegal activities. A second common divergence can be achieved by a deliberate re-calibration of the equipment to give a false reading. It is for these reasons that it is recommended that an appropriately certified patrol vessel deck officer take the fix on the boarded vessel if there is a discrepancy in readings from the position indicated just before boarding. This deck officer will be the credible witness in court, if such action becomes necessary.

Ideally, both the patrol vessel and the boarded vessel will have global positioning systems which can be cross-checked. This satellite navigating technology is very accurate and the cost of such systems is now down to a reasonable level for surveillance purposes, mid-range price is approximately $4,000 US. The FFA vessel monitoring system will link through such a system and not only provide regular position information as required in the license, but will also enable the shore station to interrogate the system on the vessel at any time for periodic, unexpected checks. This will greatly reduce the need for air surveillance for licensed vessels for position fixing and will also assist in determining the position of the vessel for boardings or any other activities when a position is required. Air surveillance, as noted earlier, will still be required to identify the activities of vessels without transponders, but this should be accomplished in a more cost effective manner due to easy identification of radar contacts without responses. Noting that all vessels may not have such technology, it is incumbent on the fisheries officer to have a working knowledge of common marine navigational systems and techniques for fixing the position of the vessel. The officer must be able to detail the procedures used in gaining the position and verification of the position of the vessel and an accurate determination of time of the fix. This will form the base for the verification of all the activities of the vessel while in the zone, and especially in the case of an alleged infraction of fisheries law.

Most larger vessels carry a variety of compasses, magnetic and gyros, radars, satellite navigation machines, omega positioning electronic equipment, sextant, direction finding equipment, echo sounders, fish finders and sonars as well as navigation publications for the area. A good general description is attached in the FFA's Fisheries Prosecution Manual, reproduced as Annex H. These pieces of navigation equipment and the observation of their use and status on boarding the vessel can be very useful in court proceedings. Examples of evidence that can be gained from observations and their use as evidence are also included in the aforementioned prosecutions manual, for example, one could note the course being steered upon first sighting the fishing vessel and check if it is the same as that given to the helmsman, the latter often being penned to a board near the wheel. The radar on a very short range, one mile or less, could indicate searching for a transponder from a fishing buoy, whereas the safety range is 10 or twelve miles. The satellite navigator often provides printouts of positions over a period and can show where the vessel has been. Direction finding equipment settings should be compared to the frequencies for the vessel's fishing buoys and cross checked with the course being steered. Echo sounders and fish finders could indicate the presence of fish being chased and prominent markings could assist in the determination of the course of the vessel over a period. The charts and logs indicate positions, times when freezers are operated, engines run and at what speeds and temperatures, etc. These observations can all assist in building and supporting a prosecution if an alleged violation has occurred. Notations as to whether the gear was warm on arrival, signifying recent use, engines functioning properly on the return to port and navigation equipment suddenly working accurately could all be points to refute claims of malfunctioning electronic equipment and the master not knowing where the vessel was in the ocean.

5.7   Transhipment

Transhipment of fish at sea is one of the most dangerous and difficult fisheries activities to monitor. It cannot be done effectively without a minimum of two persons, one on the delivering vessel and one on the receiving vessel. As the activity of two ships heaving in the sea side by side is a dangerous operation, the masters of both vessels will want to carry out the operation as quickly as possible. The fisheries officer, on the other hand, will want to ensure that there is an accurate account of the fish onboard the receiving vessel before the transfer and to have an accurate recording of the fish transferred. This will necessitate an inspection of the vessel receiving the fish before the commencement of operations and a very accurate monitoring of the fish transhipped, requiring time of both vessel masters. The difficulty in verifying the species and weight of the fish moving from one vessel to another is a challenge as this may require the officer's presence in the hold of the vessel, thus making observation from the deck of the movement of the fish impossible.

If a country refuses to permit the transhipment of fish inside its fisheries waters, then the vessels will, in all likelihood, tranship the cargo outside the zone and then apply to re-enter to continue fishing. This will result in the loss of continuity of important data on the fish removals from the zone. This can sometimes be calculated from other records, but it cannot be verified and on some occasions, it is lost from the system. It is recommended that Fisheries Administrators, in designing their MCS strategy, use negotiations to encourage the vessels to tranship their fish in their ports. This might be done through an incentive of reduced port administration costs or procedures. The country would then have both the vessels in a stable, controlled environment where the accuracy of the transfer can be monitored closely and easily.

Another international concern which can partly be addressed through the encouragement for fisheries transhipment in port is the issue of obtaining information on fisheries support vessels involved in the transhipment. These vessels often fall outside various international safety conventions as these agreements do not include "fishing vessels". On the other hand, these vessels often fall outside fisheries control mechanisms and remain unregulated. The tool of the fisheries license, supported by appropriate legislation including

68

support vessels in the definition of fishing vessels, could be a step to implementing international standards and controls for these vessels.

The requirement to license, or at least register, these vessels as part of the international fishing fleets could be an issue for further consideration. Coastal States could consider similar licensing requirements for such third party vessels and hence, require compliance with applicable laws as a pre-requisite to operating in their zone. This thereby becomes a potential component of port State control for fishing vessels. With the coming into force of the Protocol to the Torremolinos Convention, these vessels would then be required to comply with safety certification under both the regular shipping and as fishing vessels.

5.8   Port State Control and FAO "flagging agreement"

Fisheries non-compliance occurs when the economic benefits gained outweigh both the potential of detection and the penalties. If there were a system in which the risk of detection is very high and the penalties are sufficient to create a high level of deterrence, then the opportunities and practice of non-compliance of fisheries law would be minimized.

The concept of *port State control agreements* for coastal zone management, environmental protection and fisheries is receiving attention from many sectors. Currently all arrangements for port State control exclude fishing vessels, but when the Protocol to the Torremolinos comes into force, fishing vessels will be required to comply with port State safety certification. This tool is potentially very attractive in sub-regions, regions, or even bilateral situations where States can agree on the benefits which can be derived from such cooperation. At present, the International Maritime Organization and the Food and Agriculture Organization of the United Nations are both working on several initiatives which will enhance the role and potential benefits to coastal States from port State control. A paper by Fernando Plaza of IMO on the subject has noted the positive aspects and the concerns of port State control. In essence, the flag State has the responsibility of implementing the international maritime agreements to promote safety and environmental protection of the seas, but it can be through port State controls that countries maintain and monitor the current situation and ensure flag States have the appropriate information to carry out their responsibilities.

A present, there is a considerable network of port State control systems in Europe, Latin America and the Asia-Pacific region. Initiatives are commencing in the Caribbean, Southern and Eastern Mediterranean, Middle East, West and Central Africa, Eastern Africa and the Indian Ocean. Current initiatives are focused regionally to maximize the benefits and cost effectiveness of the initiatives. The activities are centred on ship safety and marine pollution controls with respect to the ILO Convention, SOLAS agreements and MARPOL agreements. There has been a need identified for standardization of training and application of the controls and a code of conduct for the control officers, but there has been a marked improvement in the control and compliance of vessels visiting ports under such a system.

The potential of port State control mechanisms to include fisheries interests with respect to port inspections, safety certification, information exchange and regional standards and cooperation for fisheries control exists, and it could be a very timely and cost effective initiative, if implemented. If renegade vessels using flags of convenience were discouraged from such practices through the potential deterrence of being detained in regional ports, then the incentive to fish in such a fashion would cease. The use of a credible central agency with appropriate sensitivity for security could greatly assist in the implementation of port State controls for fisheries MCS.

FAO has been working for several years on establishing standards for fisheries throughout the world. The most visible examples include vessel markings, gear identification and marking and, more recently, the *flagging of vessels fishing on the high seas* and the responsibility for flag State enforcement. Coupled with this are the regional fisheries initiatives to promote higher levels of compliance for all vessels in the region. The FFA regional register and agreements for information exchanges and mutual use of enforcement infrastructure have been initiatives in this direction.

The recent FAO flagging agreement, the AGREEMENT TO PROMOTE COMPLIANCE WITH INTERNATIONAL CONSERVATION AND MANAGEMENT MEASURES BY FISHING VESSELS ON THE HIGH SEAS, will, when it comes into force, assist in the promotion of conservation practices on the high seas. Flag States will be required to register all vessels authorized to carry their flags and forward this information to FAO, be responsible for the control of said vessels, and their nationals onboard to ensure they fish in a responsible manner, share information on these vessels with the international community and act on information regarding activities of its flag vessels which undermine the effectiveness of international conservation practices. Countries party to this agreement will be making a commitment to international conservation of the fisheries on the high seas. FAO will have an information database on all fishing vessels authorized by member states to fish on the high seas throughout the world. This information will be of considerable use to developing countries in deciding to license vessels to operate in their own waters. Further, regional and inter-regional cooperation on information exchange regarding vessel registrations and activity reports will enhance the potential of detection and action in the case of fishing practices which undermine international conservation principles. This initiative by FAO is a step in the right direction for international co-operation and the establishment of international standards for cooperation, conservation and a Code of Conduct for Responsible Fishing.

5.9   Fisheries Prosecutions

One of the most onerous and important tasks for a Fisheries Administrator and officers is the successful preparation and execution of a fisheries prosecution. Many fisheries offenses have resulted in acquittal in the courts due to lack of proper preparation and training by all concerned with respect to fisheries prosecutions. The inability to successfully prosecute a case in court makes the expense and effort expended on fisheries MCS activities ineffective and a considerable waste of time and money. Most recently, both the South Pacific Forum Fisheries Agency and the ASEAN countries have compiled standard manuals on prosecution procedures for their regions. One suitable reference is the recently released FFA Fisheries Prosecutions Manual provided as a guide to fisheries officers who may have to prosecute a case without having past experience in this exercise.

The success of a fisheries prosecution stems from appropriate training and preparation of all individuals involved in the case. This commences with the knowledge of the prosecutor and the judiciary of fishing, the fishing environment, the management scheme and its importance to the economy of the state and the MCS activities required for the conservation of these resources. The fisheries officer may be required to prosecute the case or it may be left to another department not familiar with fisheries activities. The education and training of these individuals has been found to be best accomplished through realistic fisheries experiences, workshops, and mock trials. This education process is best achieved, as stated in the FFA manual, *before anything happens*.

As noted in the reference, "The best way of preparing to prosecute a fisheries case is to hold an exercise".[10] The advocate, judiciary, fisheries officers, interpreters, patrol vessel crew and officials from other departments can all benefit from visiting a fishing vessel and taking part in a mock exercise in which a vessel is boarded, inspected, detained and ordered to port and charged. The moot court can also prepare all parties for the type of questions and explanations and definitions that the prosecutor will need to be able to present to the judge to ensure an understanding of the alleged infraction. The organization in place and its efficiency in addressing the details of a detained vessel, cargo and crew, which can facilitate the process considerably, can also be tested during this exercise.

Fisheries Administrators should ensure that their field patrol staff are all trained to be very observant and to note details of the situation as soon as a vessel is sighted, until it is decided to order the vessel to port or to permit it to carry on with its activities. Observations with respect to the activity on the deck of the vessel when the patrol vessel comes into view are important. Hasty activity on the deck, dumping of gear or fish, fresh fish offal in the sea, sea birds feeding, ropes or gear over the side are all indicators of fishing. Photography with time and position notations are useful in these situations. If the vessel is acting appropriately, the photographs can be used for training future officers. *The more observant the fisheries officer and accurate the notes, the easier it will be to reconstruct the events to decide whether to lay charges, which charges, and how to prosecute the case.*

The fisheries officer should always keep in mind the fact that the judiciary are not on the scene and hence will want to be able to understand events unfolding in chronological order through the explanation of the officer in court. With this in mind, the officer may wish to continually remind him/herself, "What is the judge going to ask?" or "How do I describe this?" An example of an enforcement scenario can assist in pointing out some common observations in preparing a case.

The well marked Fisheries Patrol Vessel SEA PROTECTOR from a fictitious country, called OUR COUNTRY, has received a message from a recent air patrol that there appear to be several unlicensed vessels fishing twenty miles inside the fisheries waters of OUR COUNTRY. These vessels seem to come inside to fish at night and depart the zone early in the morning. Overcast conditions and cloud have prevented the air patrol from identifying the vessels, but their speed indicates fishing operations.

---

[10] *Coventry, R.J. (1991)*

It is decided by the patrol vessel master and fisheries officer to circle the area indicated and approach the vessels in the early morning from an easterly seaward direction. This will place the patrol vessel in the sun until close identification is necessary and if the patrol vessel approaches at fishing speed the fishing vessels may not take notice until the former can get within range for identification of the vessels and activities. Sunrise is at 5:17 in the morning and the patrol vessel makes for its rendezvous with the fishing vessels at 5 in the morning. Boarding equipment is assembled, checked as ready for operations and issued during the evening. If firearms are carried, they are issued just prior to the commencement of the boarding in the morning. The boarding team and officers of the crew are briefed on the intended operation and tasks are assigned to each team member. The boarding team is assembled a half hour before sunrise, gear is checked and each team member is asked to repeat their tasks for the boarding. An officer on the bridge of the patrol vessel is assigned to note all activities during the approach and while the boarding team is on the vessel. Some of the questions the officer may note in the log could include:

1. What is the time, weather, the sea state, temperature and direction of the wind and waves?

2. What courses were the fishing vessels steering on the appearance of the patrol vessel, what speed and were there any changes on recognition of the patrol vessel?

3. Is the sea calmer around the fishing vessel, from fish oil in the water, are there any dead fish or fish offal on the sea? Are any fishing gear, buoys or small boats visible in the water or visible on the vessel? Any lines over the side, bloody water or offal in the scuppers of the vessels? Any flocks of sea birds feeding on fish scraps? Any fish activity or the surface of the sea in the immediate vicinity indicating the use of fish baiting?

4. Is the fishing vessel moving through the water, is its reduction factory working, any winches in operation, any radars operating, any communications heard on the radio? Do the patrol vessel sonars pick up any echo sounders in operations?

5. What are the deck crew doing on first sighting and is there a change in level of activity, if so in what manner, stowing gear, dumping fish, what?

6. What is the reaction of the fishing vessel on communication with the patrol vessel and orders to prepare for a boarding?

7. What are the activities seen on deck during the period when the boarding team approaches the vessel? If hostile, warn the boarding team. What actions transpire on deck during the boarding team's inspection? Has any gear been switched off during the boarding as seen from the patrol vessel?

The minimum boarding team should be comprised of four, preferably six, persons including the fisheries officer, a member from the engineering department and a ship's officer. There should also be a boarding boat operator who drops the team and stands off the fishing vessel, prepared to pick up the team. On approach to the vessel the team should observe the activity on the vessel and note the presence of increased activity, fresh fish, blood or offal, gear in a position for fishing or possibly poorly stowed, winches hooked to fishing gear, diving gear or small boats on deck and wet from recent use.

On boarding, the crew should still be observing the deck and activities of the crew. If the boarding appears to be unopposed, the fisheries officer and boarding team, save one person to remain at the head of the ladder, should proceed to meet the captain and identify themselves. Requests should be made for the fishing license, ship's log, all fishing, processing and freezer logs, and the engineering log. Two members of the team should accompany the vessel crew when retrieving these logs, if this is possible. The activities ongoing on the bridge of the vessel should be noted at the time of boarding to determine if there seems to be a flurry of activity around the navigation chart or vessel logs. The settings on the various navigational gear should be taken at this time. The inspection should then be carried out in accordance with standard procedures identified during the briefing, observing the status of fishing gear, hot from recent use, fresh fish in the freezers, fishing gear wet, blood in the production areas. Photographs are a rapid method of indicating the state of the vessel and gear on arrival on the vessel. The master is asked to indicate the position of the vessel and to respond to questions regarding the activities of the vessel.

If the inspection of the vessel and documents indicates that the master may be fishing in a closed area without authorization, the master is then ordered to take the vessel to port for further investigation. There are several opinions as to when the master should be informed that there appears to be a violation and of the appropriate legal rights available under the law. If there is an intention by the fisheries officer to ask a direct question of the master as to whether he/she knew they were fishing in a closed area, or one where they were not licensed to fish, and this will be submitted as evidence in court, then it is prudent to inform the master of rights under the law before the question is asked. On the other hand, if the officer is determining the position of the vessel and asks the master to indicate the vessel's position, there may be no need to read the rights until it is determined at a later time that a charge, or charges, may be laid. This is an important point in law and it is suggested one check how this matter is dealt with in local legal practice, with and without legal rights being read to the master. Technically, one could argue that until it is decided charges are to be laid, the boarding is an inspection only and it is assumed that the master is innocent until further review of the circumstances surrounding the events. As such, it is not necessary to read the rights to the master and unnecessarily indicate that there may be any alleged infraction of the law.

At the point in time where the master is ordered to take the vessel to port, the reaction of the master and the fishing crew is very important to the safety of the boarding party. The cooperation of the master in this process should be duly noted, as well as the performance of the vessel and its navigation and engineering gear. The MCS Central Operations Centre and appropriate port authorities should be notified of the vessel's passage to port and its expected time of arrival (ETA), so that arrangements for accommodation of the crew and security of the vessel and catch can be prepared. Preliminary documentation for court appearances can be drawn up on land and officials representing the vessel informed of the vessel's port visit.

During the passage to port the fisheries officer may wish to consider the preparation for the case. These points are very well covered in the aforementioned FFA manual (see pertinent portions in Annex H), but some of the considerations going through the fisheries officer's mind could be:

1. Is there a need for an independent expert witness(es) to check the state of the navigation gear, the freezers or the engine machinery?

2. What evidence has been gathered and how strong is it? Has it been cross checked and verified through different sources?

3. What certificates are needed to use the evidence appropriately in the case? Have all statements been taken and were warnings read to each witness when it was decided to proceed in this manner?

4. Which witnesses will be necessary to the case? When should these be interviewed? On landing and with whom present? Are interpreters needed?

5. Which charge(s) should be laid and against whom? Are they summary or indictable offenses?

6. Is all the evidence and are all exhibits secured, is there need for any more documentation? Are any other photographs necessary?

7. What is the value of the vessel, fish and penalty which can be reasonably expected from the case? Is a valuation expert needed? This will be necessary for setting a bond.

8. Can this be settled through an administrative procedure or does it need to go to court? What can happen if the fine is too high, e.g. abandon the vessel, and what follow-up procedures would then be necessary?

9. What are the procedures for setting bail for the accused and what is a reasonable amount? Where can the crew and master be accommodated?

10. Have all the notes for the case been completed for use by the prosecutor?

Armed with the answers to the above questions, the fisheries officer is ready to meet and brief the prosecutor and the Fisheries Administrator on landing in port. This briefing should be chronological and thorough and at the conclusion, the fisheries officer should have a list of recommendations of action to be taken, if there is acceptance to proceed with the case. The master should be formally charged, witness statements taken and certified, if not already done, and the crew and master accommodated. An appearance date should be set with the courts for as early as possible and the vessel, gear and fish should be secured in a manner so as not to result in spoilage of the catch. Evidence from the air patrol and statements from the air crew should be obtained and certified. The prosecutor should then review the evidence, exhibits and statements and then prepare the case for the appearance hearing.

Details for these proceedings and activities are appended, but it is sufficient at this point in time for the fisheries officer, if not the prosecutor, to remember the prosecution lays the charges and the court hears the plea of the defendant. The court then hears a summary of the case and decides, for a guilty plea the level of penalty, and for a non-guilty plea, whether there is evidence for proceeding to trial. The trial will commence with the case for the prosecution, followed by a cross examination of witnesses by the defense, the case for the defense and summary statements. This can be a lengthy process and hence the notes are doubly important to refresh one's memory of the events leading up to the laying of charges. Knowledge of international and national laws, observation, good notes and preparation will bring success to Fisheries Administrators in the prosecution of cases for serious violations of fisheries law.

Success comes from practice, experience, good fisheries law and an appropriate and professional MCS strategy and team of officers. It is hoped that this paper will assist all Fisheries Administrators in the enhancement or development of a successful MCS strategy.

REFERENCES

Allain, R.J. 1988. Monitoring, control and surveillance in selected West African coastal states with recommendations for possible ICOD interventions in the future. Ottawa, ICOD.

Allain, R.J. 1982. A study of aerial fisheries surveillance in certain CECAF coastal States. FAO CECAF/TECH/82/46 (En). 70 p. Rome, FAO.

Allen, T.D. 1991. Applications of satellite remote sensing over the Indian Ocean. Sri Lanka, IOMAC.

Armstrong, A.J. 1991. Development of International/Regional Co-operation in Fisheries Monitoring, Control and Surveillance. London, United Kingdom, Commonwealth Secretariat.

Armstrong, A.J. 1992. Survey and analysis of sub-regional MCS mechanisms in the central States of West Africa. Rome, FAO.

Bergin, A. 1988. Fisheries surveillance in the South Pacific. Ocean and Coastal Management, 1988, Vol.11, No. 6. pp. 467-491. Australia.

Bergin, A. 1993. New fishery agreements concluded in South Pacific. Ocean and Coastal Management, 1993, Vol. 19, No. 3. pp. 200-304. Australia.

Bonucci, N., Kvaran, E.R. and Palsson, O.K. 1991. An assessment of national and regional capabilities for MCS in NW Africa for a possible regional project. Rome, Reykjavik, World Bank Report.

Bradley, R. M. 1993. Fisheries sector profile of the Philippines. ADB Agriculture Department, Division 1. Manila, Philippines, ADB.

Burke, W.T. and F.T. Christy Jr. 1990. Options for the management of tuna fisheries in the Indian Ocean. FAO Fisheries Technical Paper No. 315. Rome, FAO.

Chakalall, B. (ed.). 1992. Report and proceedings of the meeting on fisheries exploitation within the exclusive economic zones of English-speaking Caribbean countries, St. George's, Grenada, 12-14 February 1992. FAO Fisheries Report No. 483, 160 p. Rome, FAO.

Chouinard, J. 1992. Propositions pour un registre sous-régional des navires de pêche de la zone Nord-Ouest Africane. FAO FI:MCS/WA/92/Inf.10 October 1992. Quebec, FAO.

Christy, L.C. 1985. Fisheries legislation in the Seychelles. Rome/Seychelles, FAO Review.

Christy, F.T. Jr. 1987. Global perspectives on fisheries management: disparities in situations, concepts and approaches. Regulation of fisheries: legal, economic and social aspects. Proceedings of a European workshop, University of Tromsø, Norway, 2-4 June 1985, Ulfstein, Anderson and Churchill (ed.) 1987, pp. 48-70. Norway, University of Tromsø.

Cirelli, M. T. 1993. Fisheries Legislation Revision and Training for Monitoring, Control and Surveillance. Report to the Government of Belize on the Revision of the Fisheries Legislation. FAO TCP/BZE/2251 (T) - FL/WECAF/93/23. 98 p. Rome, FAO.

Clark, J. R. 1992. Integrated management of coastal zones. FAO Fisheries Technical Paper 327. Rome, FAO.

Commonwealth Secretariat. 1981. Summary of a scheme of maritime surveillance and enforcement for the Solomon Islands. 10. PACEM in MARIBUS, Vienna, Austria, 27 October 1980, Proceedings and Final Report and Recommendations. pp. 137-140. United Kingdom, Commonwealth Secretariat.

Cosin, J.M. 1987. La vigilancia es una inversion. Un ejemplo pratico. The Economics of Fishing. Economia de la Pesca, Barcelona, 19-21 June, 1985, Vol.51, No. Suppl. 2. pp. 55-64. Barcelona.

Coventry, R.J. South Pacific Forum Fisheries Agency: fisheries prosecutions manual, funded by ICOD.

DFO files of January - December 1990. SWATH vessel technology and trials. Halifax, DFO.

Derham, P.J. and Christy, L.C. 1984. L'autorisation et le controle de la pêche étrangère au Madagascar. Rapport prépare pour le Gouvernement de la Republic Democratique de Madagascar par l'Organisation des Nations Unies pour l'Alimentation et l'Agriculture. FAO FL/IOR/84/12; FAO FI/GCP/INT/399/NOR; FAO FI/GCP/INT/400/NOR. 11 p. Rome, FAO.

Derham, P.J. and Christy, L.C. 1984. Licensing and control of foreign fishing in Mauritius. Report prepared for the Government of Mauritius. FAO FI/GCP/INT/399/NOR; FAO FI/GCP/INT/400/NOR; FAO FL/IOR/84/13. Rome, FAO.

Derham, P.J. 1987. The implementation and enforcement of fisheries legislation.The regulation of fisheries: legal, economic and social aspects. Proceedings of a European workshop, University of Tromsø, Norway, 2-4 June 1985, Ulfstein, Anderson and Churchill (ed.) 1987. pp. 71-81. Norway, University of Tromsø.

Derham, P.J. 1991. Proposals for the development of an Angolan fisheries protection service. FAO GCP/INT/466/NOR Document de travail 91/10 (En). Rome, FAO.

EEC Fish Reports. 1987-1991. Update on EEC agreements with ACP States to 1991. Brussels, EEC Fish Reports.

FAO. 1981. Report on an expert consultation on MCS for fisheries management. Rome, FAO.

FAO/UNDP. 1981. Report of the regional seminar on monitoring, control and surveillance of fisheries in exclusive economic zones, Jakarta, Indonesia, 30 November-4 December 1981. Manila-Philippines FAO/UNDP 1981. 166 p. Manila, Philippines, FAO/UNDP South China Sea Fisheries Development and Coordination Programme.

FAO. 1985. Fishery Information, Data and Information Service and Fishing Technology Service (comps). Definition and classification of fishery vessel types. FAO Fish. Tech. Paper. (267). 63 p. Rome, FAO

FAO. 1986. Report of the expert consultation on the technical specifications for the marking of fishing vessels. Rome, 16-20 June 1986. FAO Fish. Rep.(367): 74p. Rome, FAO.

FAO Legislative Study. 1987. A regional compendium of fisheries legislation (Indian Ocean Region), Vol. 1 - Summary, and Vol. 2 - Seychelles. Rome, FAO.

FAO. 1989. The standard specifications for the marking and identification of fishing vessels. 69 p. Rome, FAO.

FAO. 1991. Standard manual procedures of monitoring, control and surveillance of fisheries in the exclusive economic zones of ASEAN countries. 5 manuals. Jakarta, Indonesia, FAO/ASEAN.

FAO. 1992. Draft agreement for the establishment of the Indian Ocean Tuna Commission. Rome, FAO.

FAO. 1992. Report of a regional workshop on monitoring, control and surveillance for African States bordering the Atlantic Ocean (Accra, Ghana, 2-5 November 1992). FAO GCP/INT/466/NOR Field Report 92/22 (En). Rome, FAO.
(This includes country reports from participants)

FAO. 1992. Monitoring, control and surveillance of fisheries in the exclusive economic zones of Asean countries: project findings and recommendations, MCS training, completed manuals and course materials. FAO FI:DP/RAS/86/115. Rome/Jakarta, FAO.

FAO. 1993. Coastal State requirements for foreign fishing. FAO Legislative Study No.21 Rev.4, Rome, FAO.

FAO. 1993. Draft agreement on the flagging of vessels fishing on the high seas to promote compliance with internationally agreed conservation and management measures. 27th Conference of FAO, Rome, 6-25 November 1993. Document C/93/26 August 1993. Rome, FAO.

FAO. 1993. Fisheries monitoring, control and surveillance: recommendations for technical regulations in Indonesia. FAO TCP/INS 2252(A). Rome, FAO.

FAO/Directorate of Fisheries Indonesia. 1993. Fisheries monitoring, control, surveillance manual. FAO TCP/INS 2252, Jakarta, FAO.

FFA. 1993. FFA Regional legal consultation on fisheries surveillance and law enforcement, Rabaul, Papua New Guinea, 4-7 October 1993. Working papers and record of proceedings (Niue Treaty and Tonga/Tuvalu Agreement). FFA Report 93/55. Papua New Guinea, FFA.

Flewwelling, P., N. Bonucci and L.C. Christy. 1992. Revision of fisheries legislation in the Seychelles. FAO TCP/SEY/0155 L/IOR/92/30.Rev.1. Victoria - Rome, FAO.

Graham, D.C. and Booth, B.R. 1991. Fisheries monitoring, control and surveillance system in operation in Sierra Leone waters as implemented by Maritime Protection Services (Sierra Leone) Limited. National seminar on fishery industries development, 25-29 November, 1991, Freetown, Sierra Leone. Freetown, Sierra Leone.

Henderson, A.J. 1993. Draft report on MCS mission to Venezuela. Venezuela/Rome, FAO.

Huibers, H.E. 1993. Ports State view - Europe. International Conference on Flags and Quality: Dilemma or Opportunity, Netherlands, The Hague, February 1993. Netherlands, The Hague.

ICOD Project Files. 1985-1991. Seychelles fisheries projects
    - Regional MCS System for SWIO, Doucet/Surette, 1988.
    - FAO Comments on Doucet/Surette Study, D. Ardill,   February, 1989.
    - Technical Assistance to the Seychelles, 1988/89.
    - MCS Workshop, 1989.
    - WIOTO Conferences, 1990/91. Halifax, ICOD.

IOC. 1992. Guide to satellite remote sensing of the marine environment. MAN.-Guides, IOC, 1992, No. 24. 184 p. Paris, IOC.

IOFC. 1982. Report of the Workshop on Monitoring, Control and Surveillance (Mahe, Seychelles, 20-25 September 1982). Second Session of the IOFC Committee for the Development and Management of Fisheries in the Southwest Indian Ocean, Mahe, Seychelles, 13-15 December 1982. IOFC:DM/SW/82/4 December 1982. Seychelles, FAO.
(This includes the background documentation for the workshop.)

IOFC. 1982. Report of the first session of the committee for the development and management of fisheries in the Bay of Bengal, Colombo, Sri Lanka, 7-9 December 1981. FAO Fisheries Report 260. Rome, FAO.

IOFC. 1983. Report of the first session of the committee for the development and management of fisheries in the Southwest Indian Ocean, Mahe, Seychelles, 13-15 December 1982. FAO Fisheries Report 285. Rome, FAO.

IOMAC. 1987. Possible outline of a programme of action for the development of the Indian Ocean marine affairs sector (IOMAC-1/A/24) Jan.20-28, 1987. Sri Lanka, IOMAC.

IOMAC. 1987. Records of the IOMAC standing committee meetings.
First Meeting - January 28, 1987.
Second Meeting - September 7-9, 1987.
Third Meeting - November 22-24, 1988.
Fourth Meeting - July 17-21, 1989.
Fifth Meeting - May 2-5, 1990.
Sixth Meeting - September 3, 1990 (Arusha).
Seventh Meeting - July 15-19, 1991. Sri Lanka, IOMAC.

IOMAC. 1990. IOMAC information workshop and 4th Meeting on IOMAC statutes February 5-9, 1990. Jakarta, Indonesia, IOMAC.

Jennings, M.G. 1980. Mise en ouvre de la règlementation des pêches. FAO CECAF Technical Report, DAKAR, Senegal, FAO-UNDP 1980, No.80/22, CECAF/TECH/80/22 (En), Dakar, Senegal, FAO.

Kaczynski, V. 1987. Surveillance and foreign fisheries management project for the Secretary of State for Fisheries, Republic of Guinea. Seattle - Conakry, World Bank.

Kesteven, G. 1981. Monitoring of ocean systems and surveillance of uses: conceptual framework. 10. PACEM in MARIBUS, Vienna, Austria, 27 October 1980, Proceedings and Final Report and Recommendations. pp. 19-26. United Kingdom, Commonwealth Secretariat.

Keysler, Dr. H.D., Nehls, K. and Glitz, K. 1990. Draft report on internal evaluation of project status and outlook for Mauritania fisheries surveillance. GTZ Project No. 88.2278.5-01.100. Mauritania/Germany, GTZ.

Koroleva, N.D. 1989. The right of pursuit from the exclusive economic zone. 6th Anglo-Soviet Symposium on the Law of the Sea and International Shipping. Marine Policy, 1990 Vol.14, No.2. pp. 137-141. London.

Kuruc, M. 1993. The Lacey Act. Organization for Economic Co-operation and Development, Ad hoc Expert Group on Fisheries, 6th Session of the Workshop on Enforcement Measures, Paris, 20 September, 1993. Paris, OECD.

Lane, P. and Associates Limited. 1991. Legal enforcement in events of maritime pollution incidents from shipping within the Caribbean Region and neighbouring international waters: A Latin American perspective. Paper presented to the First International Conference on the Legal Protection of the Environment. Havana, Cuba.

LUX-Development. 1993. Reports from the sub-regional workshop on the appropriateness and effectiveness of aerial surveillance, 8-10 November, 1993, Bangul, Gambia. LUX-Development Project AFR 009. LUX-Development.
(These papers include submissions from Sierra Leone, Guinea-Bissau, Guinea, Gambia.)

LUX-Development. 1993. The Gambia, monitoring, control and surveillance operations manual. LUX-Development Project AFR 009. Gambia, LUX-Development.

Majid, S.B.A. 1985. Controlling fishing effort: Malaysia's experience and problems. FAO Fisheries Report 289(3). pp. 319-327. Rome, FAO.

Massin, J.M. 1984. National requirements for airborne maritime surveillance. Remote Sensing For The Control Of Marine Pollution, NATO Challenges Modern Society, Vol.6. pp. 27-41.

Mc Clurg, T. 1993. Two fisheries enforcement paradigms: New Zealand before and after ITQs. Organization for Economic Co-operation and Development, Ad hoc Expert Group on Fisheries, 6th Session of the Workshop on Enforcement Measures, Paris, 20 September, 1993. Paris, OECD.

Monsaingeon. 1991. Monitoring fishing activity using the Argos system. Magazine article from ARGOS, December 1991.

Moore, G.K.F., H. Walters and D. Robin. 1991. Implementation of harmonized fisheries legislation in the OECS region. FAO Study. Rome, FAO.

Moore, G. 1993. Enforcement without force: new techniques in compliance control for foreign fishing operations based on regional cooperation. Ocean Development and International Law, Volume 24, pp.197-204. Edited version of a paper presented at the 26th Annual Conference of the Law of the Sea Institute held in Genoa, Italy, 22-26 June, 1992, United Kingdom, Law of the Sea Institute.

Munita-Ortiz, C. 1992. Sistema de vigilancia de pesca por satelite. Chile-Pesq., 192, No. 70, pp. 53-55. Chile.

Nedelec, C. and Prado, J. 1990. Definition and classification of fishing gear categories. FAO Fisheries Technical Paper No. 222, Revision 1. 92 p. Rome, FAO.

Newton, C.H. 1982. Assistance to the ASEAN countries in monitoring, control and surveillance of fisheries in exclusive economic zones. FAO Mission Report to ASEAN Countries, December 1982. Rome, FAO.

Newton, C.H. 1985. Control over fisheries in the exclusive economic zone of the Republic of the Cape Verde Islands. FAO GCP/INT 399/NOR Technical Report UNDP/FAO-CVI/82/003. Praia, Cape Verde Islands.

Newton, C. and Christy, L. 1985. Report to the People's Democratic Republic of Yemen: assessment of the requirements for the monitoring, control and surveillance of fisheries. FAO FI/GCP/INT/399/NOR, FI/GCP/INT/400/NOR, FL/IOR/85/18. Rome, FAO.

Newton, C.H. 1989. Monitoring, control and surveillance of fisheries in exclusive economic zones. RSC Series No. 49. pp. 209-213. Rome, FAO.

Okamoto, J. 1993. Introduction of the outline of the projects to prevent the poaching of fish. Organization for Economic Co-operation and Development, Ad hoc Expert Group on Fisheries, 6th Session of the Workshop on Enforcement Measures, Paris, 20 September, 1993. Paris, OECD.

O'Reilly, A. and Clarke, K. 1993. Training and needs analysis. Final report. **Fisheries report series.** pp. 116. Belize, CARICOM Fisheries Resource Assessment and Management Program.

Palmason, S.R. 1993. Supervision of the utilization of fishery resources off Iceland. Organization for Economic Co-operation and Development, Ad hoc Expert Group on Fisheries, 6th Session of the Workshop on Enforcement Measures, Paris, 20 September, 1993. Paris, OECD.

Ramos, S. 1993. Cap Vert, propositions pour l'establissement d'un systeme de suivi, controle et surveillance des peches. FAO TCP/CVI/2252. Rome, FAO.

Rao, N.S. 1986. Poaching of fish by alien vessels: need for a national strategy. Seafood-Export-J. 1986, Vol.18, No.9. pp.15-20.

Savini, M. & B.H. Dubner. 1979. Legal and institutional aspects of fisheries management and development in the exclusive economic zone of the Republic of the Seychelles. Technical Report Indian Ocean Programme, (3): 94p. Rome, FAO.

Schowengardt, L.N. Jr. 1992. Report prepared for the Government of Myanmar on monitoring, control and surveillance of the fisheries within the exclusive economic zone. FAO GCP/INT/466/NOR Field Report 92/18. 13 p. Rome, FAO.

Schowengerdt, L.N., Jr. 1985. Enforcement of foreign fishing quota allocations by the United States Coast Guard. FAO Fisheries Report 289 (3). pp. 283-290. Rome, FAO.

Senoo, H. 1983. Some problems with respect to the enforcement of judicial police power in the exclusive economic zone. Exclusive Economic Zone. Proceedings of the 7th International Ocean Symposium, October 21-22, 1982. pp. 53-54. Tokyo, Japan, The Japan Shipping Club, Tokyo, Japan, Ocean Association of Japan.

Sevaly Sen. 1989. EEC fisheries agreements with ACP States and their likely impact on artisanal fisheries (draft).

Seychelles Fishing Authority. 1987. Regional cooperation in tuna management workshop on monitoring and control of the Seychelles exclusive economic zone. FAO Project (RAF/87/008). Rome, FAO.

Survival Systems Limited. 1987. Maritime monitoring, surveillance and control, search and rescue within the exclusive economic zone of the Republic of the Maldives. Male, Maldives, ICOD Report.

Sutinen, J.G. 1988. Enforcement economics in exclusive economic zones. GEOJOURNAL, 1988, Vol.16, No.3, Special Issue "Marine Resource Economics". pp. 273-281. Rhode Island, USA.

Sutinen, J.G. and Gauvin, J.R. 1989. Assessing compliance with fisheries regulations. Maritimes, 1989, Vol. 33, No. 1. pp. 10-12. Massachusetts, USA.

Sutinen, J.G. 1993. A framework for the study on the efficient management of living marine resources. Organization for Economic Co-operation and Development, Ad hoc Expert Group on Fisheries, 6th Session of the Workshop on Enforcement Measures, Paris, 20 September, 1993. Paris, OECD.

Tanigushi, C. 1984. Jurisdiction, enforcement and dispute settlement in the Law of the Sea Convention. Consensus and Confrontation: The United States and the Law of the Sea Convention, Van Dyke, J.M. (ed.), 1984, no. 85-01 pp. 463-482.

Van Helvoort, G. 1986. Observer program operations manual. FAO Fisheries Technical Paper 275. Rome, FAO.

Van Houtte, A.R., N. Bonucci and W.R. Edeson. 1989. A preliminary review of selected legislation governing aquaculture. UNDP/FAO Study. Rome, FAO.

Anon. 1993. Articles on remote sensing of fisheries management. Mobile Monitoring Newsletter, Number 4, January 1993. Paris.

Anon. 1993. International conference on flags and quality: dilemma or opportunity. Memorandum of Understanding in Port State Control, Netherlands, The Hague, February 1993. Netherlands, The Hague.

<u>FISHING VESSEL IDENTIFICATION AND MARKING</u>

The rapid identification of a vessel type and its identification greatly facilitates MCS activities. The efforts of FAO towards standardizing the vessel markings to correspond with international radio call signs is an added advantage for identification and initiating communications with the sighted vessel.

This appendix has been provided courtesy of FAO to enable Fisheries Administrators to have key MCS information in one reference paper. The contents of this annex are reprints from FAO Fisheries Technical Paper, Definition and Classification of Fishery Vessel Types, 267 and The Standard Specifications for the Marking and Identification of Fishing Vessels. The original publications provide more detailed information in English, French and Spanish.

The first section of this annex addresses the fishing vessel marking requirements designed through several years of work by FAO.

THE STANDARD SPECIFICATIONS FOR THE MARKING AND IDENTIFICATION OF FISHING VESSELS

FOREWORD

The need for an international standard system for the marking and identification of fishing vessels was included in the Strategy for Fisheries Management and Development approved by the 1984 FAO World Fisheries Conference. An Expert Consultation on the Marking of Fishing Vessels convened by the Government of Canada, in collaboration with FAO, in Halifax, Nova Scotia, Canada, March 1985, elaborated the basis for a standard system.

A review of the report of this Expert Consultation by the Sixteenth Session of the FAO Committee on Fisheries resulted in a further Expert Consultation on the Technical Specifications for the Marking of Fishing Vessels convened in Rome, June 1986.

The Specifications contained herein were endorsed by the Eighteenth Session of the FAO Committee on Fisheries, Rome, April 1989, for adoption by States on a voluntary basis as a standard system to identify fishing vessels operating, or likely to operate, in waters of States other than those of the flag State. The Director-General of FAO has informed the Secretary-Generals of the International Maritime Organization (IMO) and the International Telecommunication Union (ITU) of the adoption of these Standard Specifications as an aid to fisheries management and safety at sea.

1. GENERAL PROVISIONS

1.1 <u>Purpose and scope</u>

1.1.1 As an aid to fisheries management and safety at sea, fishing vessels should be appropriately marked for their identification on the basis of the International Telecommunication Union Radio Call Signs (IRCS) system.

1.1.2   For the purpose of these Standard Specifications, the use of the word "vessel" refers to any vessel intending to fish or engaged in fishing or ancillary activities, operating, or likely to operate, in waters of States other than those of the flag State.

1.2   Definitions

For the purpose of these Specifications:

i) the word "vessel" also includes a boat, skiff or craft (excluding aircraft) carried on board another vessel and required for fishing operations;

ii) a deck is any surface lying in the horizontal plane, including the top of the wheelhouse;

iii) a radio station is one that is assigned an International Telecommunication Union Radio Call Sign.

1.3   Basis for the Standard Specifications

The basis for the Standard Specifications, the IRCS system, meets the following requirements:

i) the use of an established international system from which the identity and nationality of vessels can be readily determined, irrespective of size and tonnage, and for which a register is maintained;

ii) it is without prejudice to international conventions, national or bilateral practices;

iii) implementatjon and maintenance will be at minimum cost to governments and vessel owners; and

iv) it facilitates search and rescue operations.

2. BASIC SYSTEM AND APPLICATION

2.1   Basic system

2.1.1 The Standard Specifications are based on:

i) the International Telecommunication Union's system for the allocation of call signs to countries for ship stations; and

ii) generally accepted design standards for lettering and numbering.

2.1.2 Vessels shall be marked with their International Telecommunication Union Radio Call Signs (IRCS).

2.1.3 Except as provided for in paragraph 2.2.6 below, vessels to which an IRCS has not been assigned shall be marked with the characters allocated by the International Telecommunication Union (ITU) to the flag State (see pages 25 to 28@ and followed by, as appropriate, the licence or registration number assigned by the flag State. In such cases, a hyphen shall be placed between the nationality identification characters and the licence or registration number identifying the vessel.

2.1.4 In order to avoid confusion with the letters **I** and **O**, it is recommended that the numbers **1** and **0**, which are specifically excluded from the ITU call signs, be avoided by national authorities when allocating licence or registration numbers.

2.1.5 Apart from the vessel's name or identification mark and the port of registry as required by international practice or national legislation, the marking system as specified shall, in order to avoid confusion, be the only other vessel identification mark consisting of letters and numbers to be painted on the hull or superstructure.

2.2   Application

2.2.1 The markings shall be prominently displayed at all times:

i) on the vessel's side or superstructure, port and starboard; fixtures inclined at an angle to the vessel's side or superstructure would be considered as suitable provided that the angle of inclination would not prevent sighting of the sign from another vessel or from the air:

ii) on a deck, except as provided for in paragraph 2.2.4 below. Should an awning or other temporary cover be placed so as to obscure the mark on a deck, the awning or cover shall also be marked. These marks should be placed athwartships with the top of the numbers or letters towards the bow.

2.2.2 Marks should be placed as high as possible above the waterline on both sides. Such parts of the hull as the bow and the stern shall be avoided.

2.2.3 The marks shall:

i) be so placed that they are not obscured by the fishing gear whether it is stowed or in use;

ii) be clear of flow from scuppers or overboard discharges including areas which might be prone to damage or discolouration from the catch of certain types of species; and

iii) not extend below the waterline.

2.2.4 Undecked vessels shall not be required to display the markings on a horizontal surface. However, owners should be encouraged, where practical, to fit a board on which the markings may be clearly seen from the air.

2.2.5 Vessels fitted with sails may display the markings on the sail in addition to the hull.

2.2.6 Boats, skiffs and craft carried by the vessel for fishing operations shall bear the same mark as the vessel concerned.

2.2.7 Examples of the placement of marks are set out in pages 47 to 69.

3.   TECHNICAL SPECIEICATIONS

3.1   Specifications of letters and numbers

3.1.1 Block lettering and numbering shall be used throughout.

3.1.2 The width of the letters and numbers shall be in proportion to the height as set out later in this paper.

3.1.3 The height (h) of the letters and numbers shall be in proportion to the size of the vessel in accordance with the following:

a) for marks to be placed on the hull, superstructure and/or inclined surfaces:

Length of vessel overall (LOA) in meters (m)	Height of letters and numbers in meters (m) to be not less than
25 m and over	1.0 m
20 m but less than 25 m	0.8 m
15 m but less than 20 m	0.6 m
12 m but less than 15 m	0.4 m
5 m but less than 12 m	0.3 m
Under 5 m	0.1 m

b) for marks to be placed on deck: the height shall not be less than 0.3 m for all classes of vessels of 5 m and over.

3.1.4 The length of the hyphen shall be half the height of the letters and numbers.

3.1.5 The width of the stroke for all letters, numbers and the hyphen shall be h/6.

3.1.6 Spacing:
   i) the space between letters and/or numbers shall not exceed h/4 nor be less than h/6;

ii) the space between adjacent letters having sloping sides shall not exceed h/8 nor be less than h/10, for example A V.

3.2   Painting

3.2.1 The marks shall be:

i) white on a black background; or

ii) black on a white background.

3.2.2 The background shall extend to provide a border around the mark of not less than h/6.

3.2.3 Good quality marine paints to be used throughout.

3.2.4 The use of retro-reflective or heat-generating substances shall be accepted, provided that the mark meets the requirements of these Standard Specifications.

3.2.5 The marks and the background shall be maintained in good condition at all times.

4.   REGISTRATION OF MARKS

4.1 The International Telecommunication Union maintains and updates a worldwide register of International Radio Call Signs that contains details of the nationality of the vessel and its name.

4.2 In addition to maintaining a separate register of its vessels to which IRCS have been assigned, the flag State shall also maintain a record of vessels to which it has given a nationality identifier (allocated by the ITU) followed by the hyphen and licence/registration number; such records should include details of the vessels and owners.

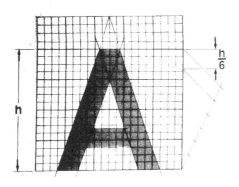

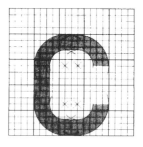

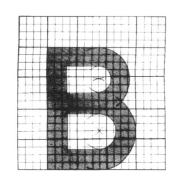

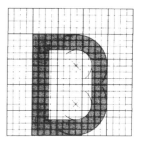

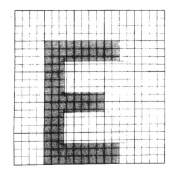

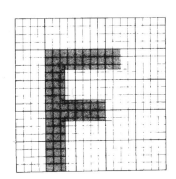

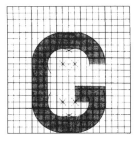

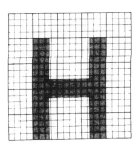

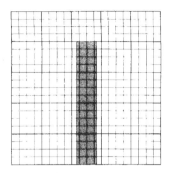

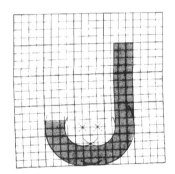

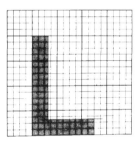

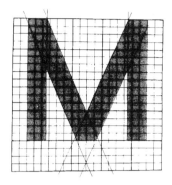

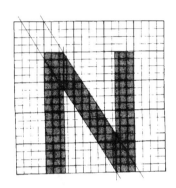

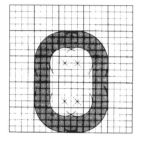

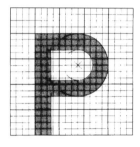

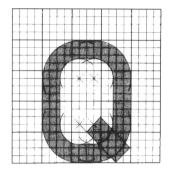

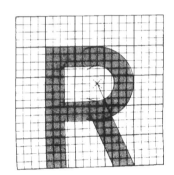

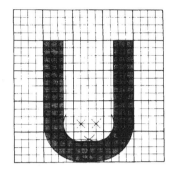

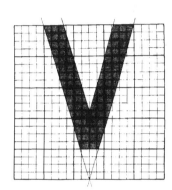

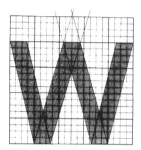

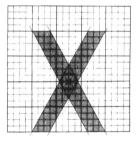

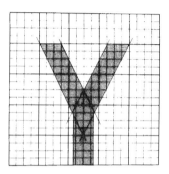

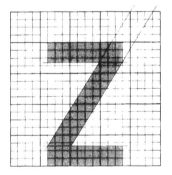

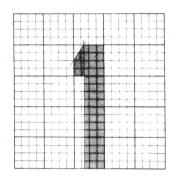

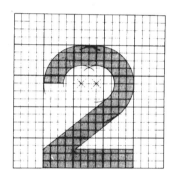

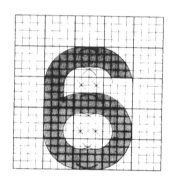

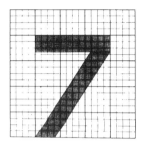

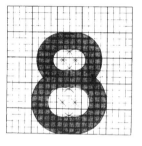

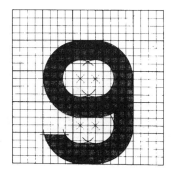

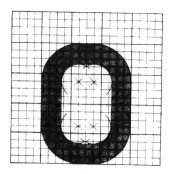

98

Examples of placement of the marks
Exemples d'emplacement des marques
Ejemplos de colocación de las marcas

CONTRAST / CONTRASTE / CONTRASTE

$\frac{h}{6}$

$\frac{h}{6}$

COLOURED BACKGROUND / FOND COLORE / FONDO EN COLOR

VESSELS AND MARKINGS

Mothership with catchers

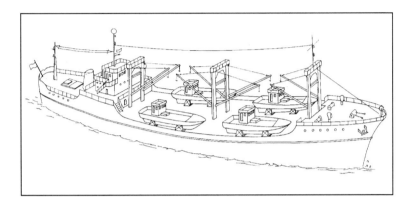

Factory Mothership

TRAWLERS

Side Trawler - 51 m

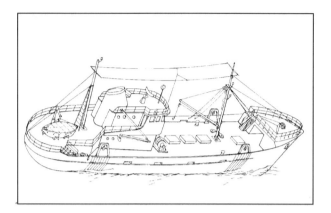

Trawling

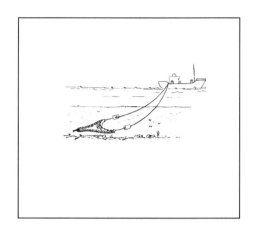

Freezer Factory Trawler - 56m

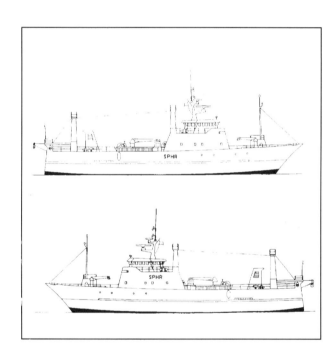

Trawling

Small Stern Trawler - 13m

Mid-Sized Stern Trawler -22m

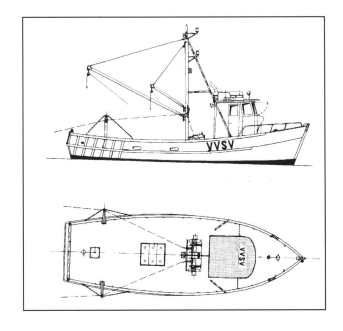

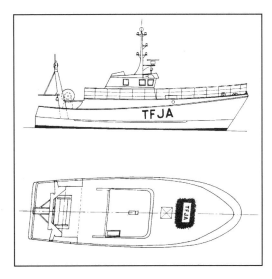

Beamtrawler - 40m

Trawler/Seiner - 20m

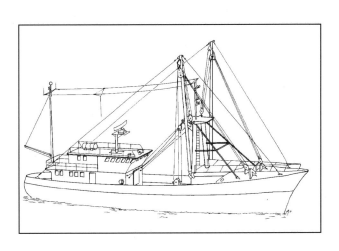

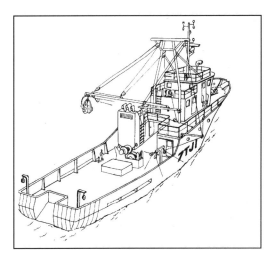

Outrigger Trawler - 18m

Outrigged and Trawling

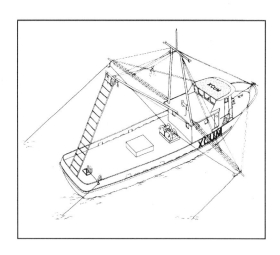

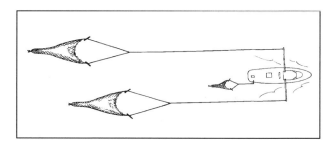

SEINERS

European Purse Seiner - 29 m

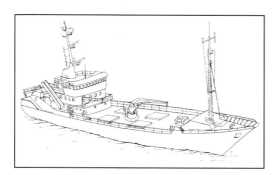

Seining

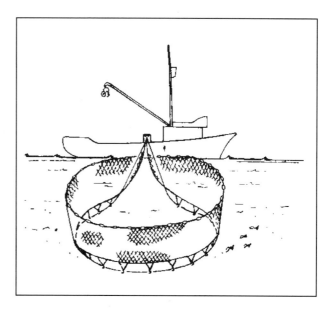

Tuna Purse Seiner - 64 m

Pursing the net

Small Purse Seiner

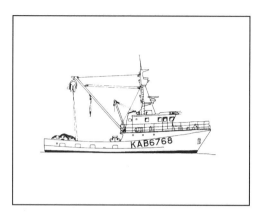

Seine Netters - 16 m

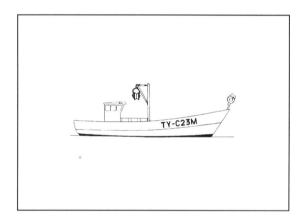

Seine Netting

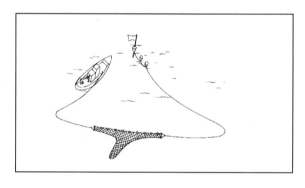

LONGLINERS

Large Longliner - 33 m

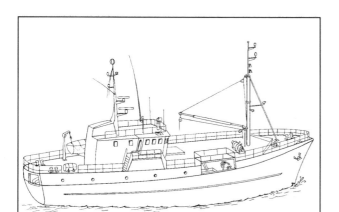

Small Longliner - 14 m

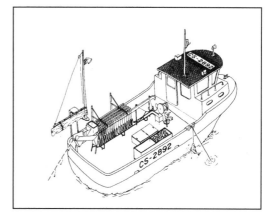

Tuna Longliner - 66 m

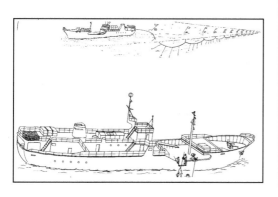

Dredger - 22 m

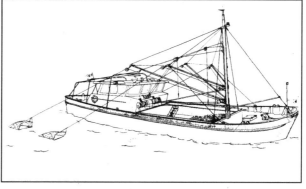

OTHERS

Troller - 16.8 m   **Trolling**

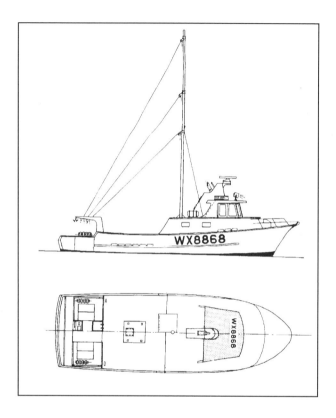
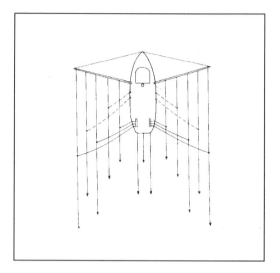

Pumper - 13 m   **Japanese Pole and Line - 38 m**

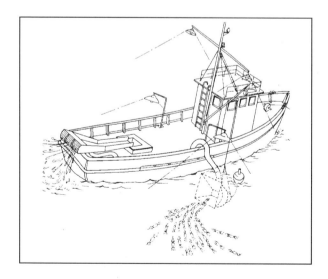
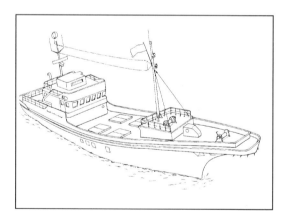

American Pole and Line - 34 m

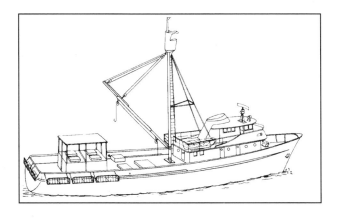

Multi-purpose Vessel - 9 m

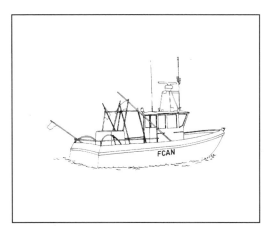

Large Pot Vessel - 26 m

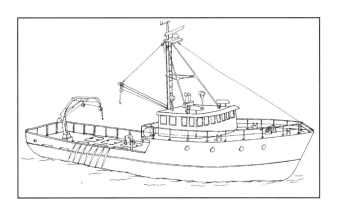

Small Pot Vessel - 6 m

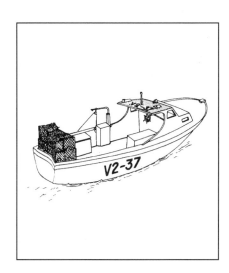

Liftnetter - 45 m

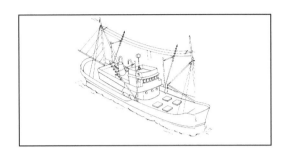

Liftnet Fishing

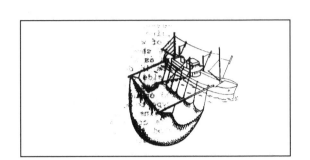

Sail - 7.4 m

Gill netting

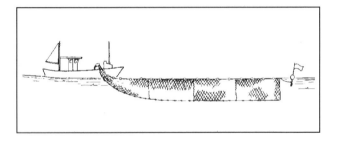

Handliner

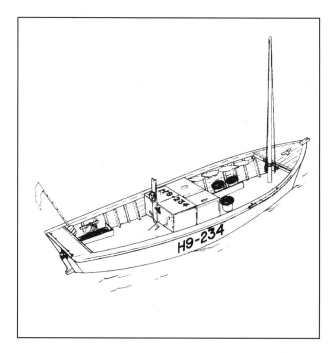

Outboard - 4.8 m

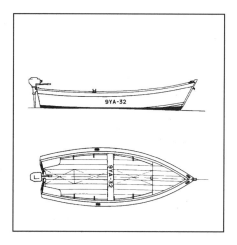

FISHING GEAR IDENTIFICATION

All Fisheries Administrators and their staff have a need to be able to identify fishing apparatus and have knowledge as to how it entraps fish. This annex provides basic knowledge of the various fishing gear types in use in the world today.

Fisheries Officers come upon fishing gear during their patrols. In the case of gear which is set illegally, it is advantageous to be able to identify the owner of the fishing gear for further discussions. In the case of legal gear, it may also be necessary to identify the owner. Many fisheries laws now require fishers to mark their fishing gear with tags in a prominent part of the apparatus where it is easily seen. The markings are often the same as required for vessels, the call sign or name of the owner.

This annex has been provided courtesy of FAO to enable Fisheries Administrators to have key MCS information in one reference paper. This annex draws heavily from FAO Fisheries Technical Paper, Definition and classification of fishing gear categories, 222 Rev.1. The original publication provides greater detail in English, French and Spanish. The division of the diagrams of the fishing gear follows the same sequence as the original publication.

SURROUNDING NETS

These nets surround the fish on the sides and extend underneath so the fish cannot escape. Purse seines are surrounding nets which after being set can be pulled together at the bottom, closing like a purse, and thus trapping the fish in the net.

<u>Purse Seine</u>

Lampara Net

This is a surrounding net without a purse line. It does have a scoop like a spoon as seen in the following diagram. The ring net is more like a purse seine with bridles to help pull in the net.

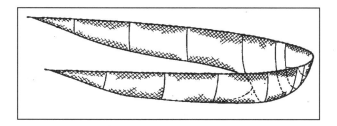

SEINE NETS

Beach Seine

These nets are usually set from a boat, but can be hauled from shore or the boat itself. The intent is to surround a large area of water to trap the fish in the area. As the net is hauled in the fish come ahead of it. Some beach seines have a bag in the centre to assist in driving the fish to the centre of the net as it is drawn to shore.

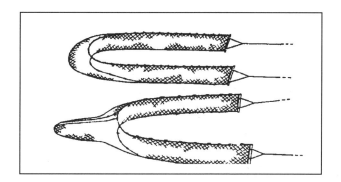

Boat Seines

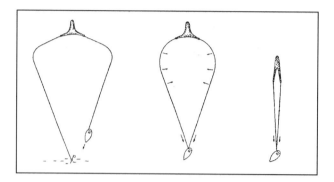

TRAWL NETS

These are towed nets consisting of a cone-shaped body, closed by a bag or codend and extended at the opening by wings. They can be towed by one or two vessels and different nets are used for bottom and mid-water trawling. In certain cases they can be rigged to sit off the side of the vessel (outrigger), or multi-rigged with more than one net being towed at the same time.

Bottom Otter Trawls

These nets operate near or on the bottom. They are held open by a ballasted ground rope coupled with float bobbins for the head rope. The design of the net is specific to the fish being sought, low opening for demersal fish and higher opening nets for semi-demersal and pelagic species.

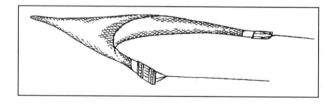

Beam Trawls

These are operated in a similar manner to the bottom otter trawls.

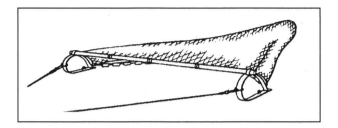

Bottom Pair Trawls

This is one large net towed by two vessels.

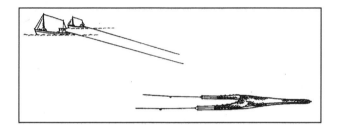

Midwater Trawls

These nets are much larger than bottom trawls. They herd the pelagic and surface fishes towards the after end of the net. Their depth is usually controlled by means of a net sounder. They may be towed by one or two vessels.

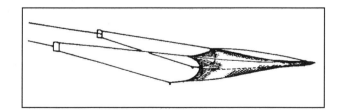

Midwater Pair Trawls

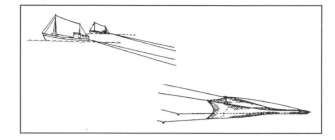

Twin Otter Trawls

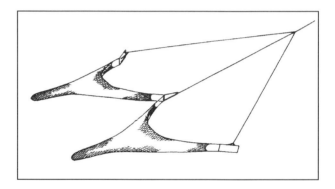

DREDGES

These are gear dragged along the bottom, usually to collect molluscs such as mussels, oysters, scallops and clams. The catch is held in a sort of bag or sieve which allows water, sand and mud to run out.

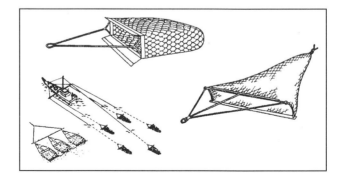

LIFT NETS

These nets are set in such a manner as to allow the fisher to attract fish with lights or bait. When they are over the net it is raised or hauled in to capture the fish. Lift nets come in various shapes and sizes. The two examples shown are for boats and smaller shore mounted apparatus.

Boat Lift Nets

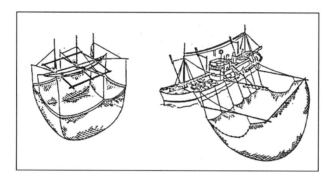

Shallow Lift Nets

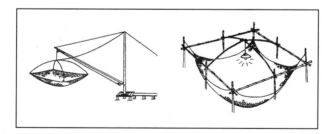

FALLING GEAR

Cast Nets

These nets are operated from boats or shore. The net falls on the fish thus trapping them against the bottom of the sea.

Other Falling Gear

Cover pots and lantern nets operated in very shallow water fall in this category.

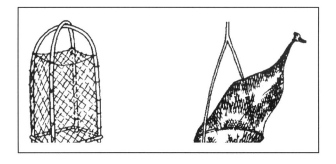

GILLNETS AND ENTANGLING NETS

These nets are used to enmesh, or catch the fish by the gills, entangling them in the net itself. Different types of nets can be used together in one gear and they may be set in long lines, called "fleets". These nets can be set at any depth and can drift or remain fixed to the sea bottom.

Set Gillnet

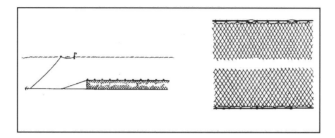

Drifting Gillnets

These nets drift freely at their set depth, on, or near, the surface.

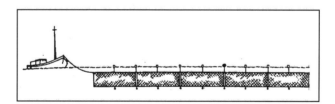

Encircling Gillnets

This gear is set with the floats on the surface and the fish are circled in shallow water. Noise or other means are used to force the fish to gill themselves in the surrounding netting.

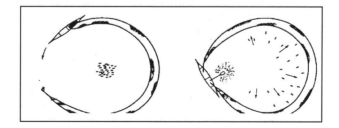

Fixed Gillnets (on stakes)

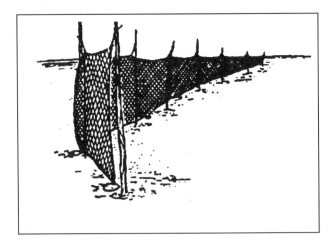

Trammel Nets

These bottom set nets are made with three walls of net, the two outer walls are larger mesh than the loose, smaller mesh centre wall. This entraps the fish in the inner wall after passing through the outer walls.

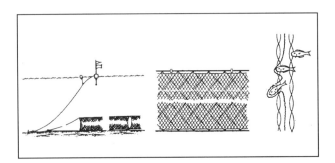

Combined Gillnets-Trammel Nets

The bottom of this gear is trammel net to catch bottom fish and the upper netting is regular gillnetting to enmesh semi-demersal and pelagic species.

TRAPS

Stationary Uncovered Pound Nets

These are large fixed nets, open to the surface, with various herding devices to retain the fish in the final "room" which is often closed at the bottom with netting.

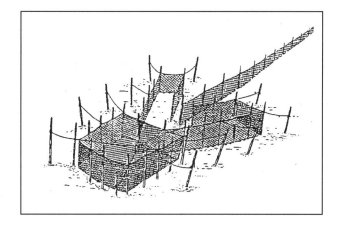

Pots

These traps are used to catch fish or crustaceans by using cages with, or without, baited interiors. They can be set singly, or in strings.

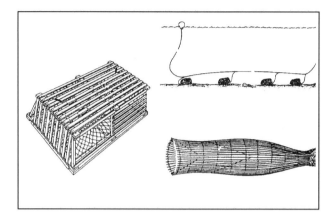

Fyke Nets

These nets are set in shallow water and may be set separately or in groups.

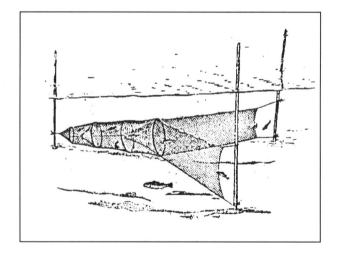

Stow Nets

These are riverine nets for use in strong currents. They are anchored in the current and the framed openings face the current.

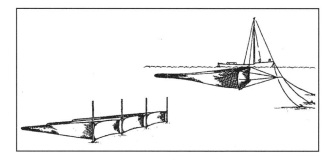

Barriers, Fences, Weirs, Corrals

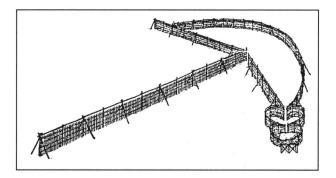

Aerial Traps

This gear is to trap jumping fish. They come to a barrier and are sometimes frightened to make them jump onto the "veranda net" set on the surface.

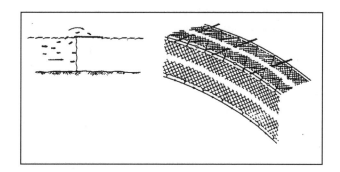

HOOKS AND LINES

Some fish are attracted to natural or artificial bait on a hook. There are many arrangements which can be constructed to catch fish in this manner with either single hooks or in a series. Some fish are attracted to hooks and then "jigged" when the hooks are hauled up and down in jerky movements. This is the principle behind the attraction of squid to the jigs on which they are caught.

Handlines and Poles

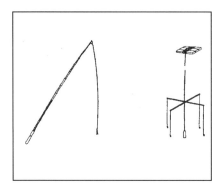
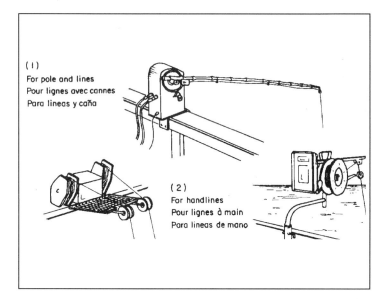

Set Longlines

These longlines have a main line to which "snoods", smaller hooked lines, are attached and baited. These snoods are set at fixed intervals and the line can then be set at any depth in the sea. These lines may be set vertically in the sea.

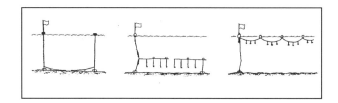

Drifting Longlines

These lines are usually set near the surface. Drifting longlines may be very long with some tuna longlines known to be 100 Km in length.

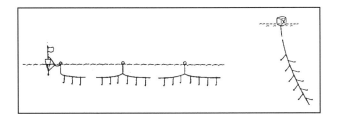

Trolling Lines

Several lines with natural or artificial bait are trailed behind a vessel to attract fish.

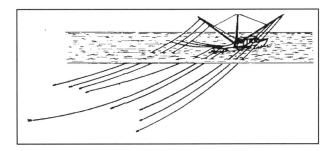

OTHERS

This section covers several gear types for which there are no diagrams. These include harpoons, spears, arrows, prongs, tongs, clamps and various scoop nets, hand implements used for fishing, poisons, explosives and electrical fishing. The two last gears are the pumper and mechanical dredges. These are methods of extracting fish and molluscs from the sea.

Pumper

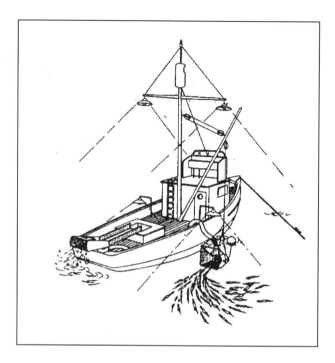

Mechanized Dredges

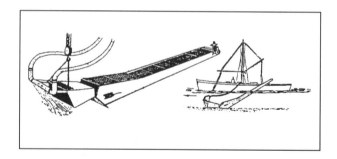

CORE COMPONENTS OF A FISHERIES OFFICER'S

OPERATIONS MANUAL

The fact that the MCS system for each country is going to be unique has been stressed throughout this paper, but it is true that there are core components and subject matter that should be included in every fisheries officer's operations manual(s). The actual content of each section will show the variability as to how each state wishes to address the subject matter. There is no one example that will be fully correct for each state consequently, it is recommended that each Fisheries Administrator use this annex as a guide only as to the subject matter for a manual.

A fisheries manual can be comprised of one or several documents which, as a whole, form the directives for fisheries officers. As the fisheries management and MCS procedures will evolve over time to address the changing situations encountered, the manual system should be easily amended. Records of the amendments should be included in each document, as this procedure then permits the reader to know under which latest authorities they are operating.

A 1991 initiative in the ASEAN countries resulted in a series of five manuals for general guidance in patrolling fisheries waters. These manuals were titled:

MCS I: Conduct of MCS Officer in Patrolling Fisheries Waters
MCS II: Procedure to Plan An Operation
MCS III: MCS Radio Communications
MCS IV: Log Books
MCS V: Guidelines on Prosecution

The Member States of the Forum Fisheries Agency have developed, over time, minimum terms and conditions for fishing in their collective fisheries zones. Training in MCS activities is regionally executed and, consequently, common standards for operations are evolving. The most recent document for MCS has been the aforementioned Fisheries Prosecutions Manual with two other papers expected in the near future, one on vessel monitoring systems and their use and a second on common terms for USE OF FORCE in fisheries MCS activities.

There are several examples of common fisheries operational procedures such as the regional system for the Northwest Atlantic Fisheries Organization (NAFO) Conservation and Enforcement Measures.

A new publication, which may be available in the near future, promises to provide a very detailed explanation on the definition of coastal and offshore zones, fishing agreements, terms and the negotiations for such with international fishing interests, use and procedures for surveillance, observers and the administrative requirements and procedures behind these operations. This document could assist many Fisheries

Administrators in further addressing the development of their operations manuals.

Common to most of these manuals, both on a national and regional basis, is an introductory section which sets out the organization, its mandate and policies with respect to fisheries management. This introduction now often includes direction as to the interaction expected between fisheries and other coastal and ocean interests of the government(s). In the Philippines, for example, there is an inter-agency committee on enforcement which addresses fisheries and other coastal zone management concerns. This committee provides overall direction to ministries involved in MCS activities.

The optimal National MCS organization, according to a consultant for the Commonwealth Secretariat in 1992, included an office of Executive Direction (President/Prime Minister and Cabinet), a Policy and Coordination Committee (Lead Minister for MCS and concerned Public and Private Officials), Chief of MCS (Officer in Charge of day-to-day operations), Surveillance Centre (Officers for coordination of specific operations) and Information Data Collection and Compilation. The latter includes the information from ship-based personnel and equipment, intelligence networks, aircraft, other countries, reporting systems and personnel on MCS activities from other agencies.

Following the introductory section might be a reasonable place to include the authorities and powers vested in fisheries officers. This then sets the stage for officers as to their duties and responsibilities. The identification for fishery officers is often referred to in this section with appropriate descriptions in appendices. Examples of identification cards follow:

FFA Identification Card

FFA VESSEL FLAG

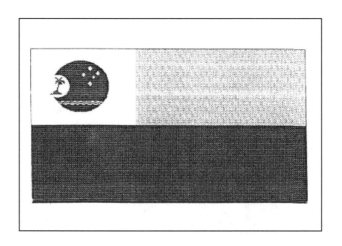

NAFO IDENTIFICATION CARD

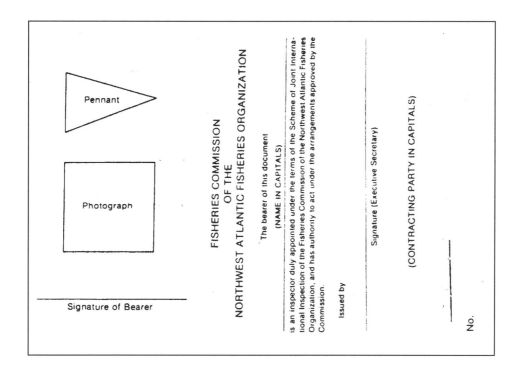

NAFO VESSEL FLAG

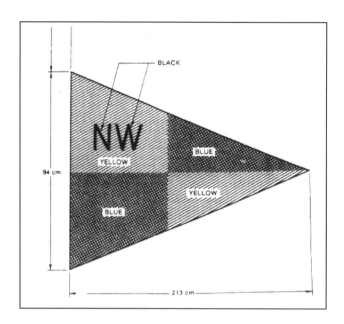

A common section included in all manuals is the current fisheries management plans for each fishery. Accompanying each of these plans should be the operational strategy to implement the plan. It is this latter document which is of considerable importance for the fisheries officer, as it sets not only the priorities for action, but also the detailed procedures to be taken for monitoring, the control directives and the surveillance/enforcement action. Some of these procedures are common and hence can be grouped, but any special considerations for each fishery are best noted along side the management plan and implementing policy. These plans would provide information on the management system for the fishery; effort, area, overall quotas, individual quotas, trip limits and others noted earlier. The fishing gear permitted and its attachments or prohibitions on methods for setting would also be noted here. The MCS plan would detail how the different aspects of the plan are to be monitored and surveilled, if special techniques are required. Special monitoring requirements for data would be noted in this section as well, including any special obligations for the fishers. This procedure would be necessary for each fishery, and if there were different requirements for sectors of each fishery, due to agreements with the industry or international negotiations, fisheries officers would need to be familiar with these conditions as well.

There are other core sections for each manual which would not change appreciably over time. These include the gear specifications, measuring methods, approved attachments, markings for vessel and gear identification, reports required from the fishers, internationally recognized instructions for stopping the vessel for inspections, boarding procedures, arrest procedures, the international aspects of the Convention onthe Law of the Sea, pertinent regulations, communications procedures and codes, to name a few. The boarding and inspection procedures, the approach to the fisher for monitoring are procedures with which every fisheries officer should be very familiar. Approved reaction to hostile and aggressive responsesand to inspection and monitoring activities should be clearly described in the fisheries officer's manual. This relates closely to the powers and authorities and will impact significantly on the success of prosecution activities when these are necessary. This section, commencing with the identification of the officer(s), through the entire inspection or monitoring procedure to the final reports and follow-up action, should be detailed in the manual and reference to it included in every briefing of officers.

Other information for officers could include fish identification guides, gear identification guides and monitoring/measurement requirements for fisheries stock assessment activities.

Therefore, without pre-empting the prerogative of the Fisheries Administrator with respect to format and appendices, a possible operations MCS manual could resemble the following:

FISHERIES OFFICER OPERATIONAL MANUAL

Amendment List and Dates

Introduction

    Background and history of the fisheries in the country.
    Organization
    Mandate
    Linkages with other Ministries
    Linkages with other Governments for support

Authorities and Powers

    Identification of Fisheries Officers and their equipment,   vessels, vehicles and aircraft.
    Acts
    Regulations
    Fisheries Agreements   - national   joint ventures, special fishing for
                                                  research, etc.
                                  - international
    Convention on the Law of the Sea -   pertinent clauses and definitions of
                                                  zones

Fisheries Management Plans
    These would include all the parameters for each plan including the controls to be used, effort, quotas, areas, gear, seasons and their combinations.

Plan 1 -   Pelagic
                Fish Identification Guides
                National Plan
                International Plan
                MCS Implementation Plan
                      data collection
                      special regulations and policies
                      enforcement strategy

Plan 2 -   Demersal
                Fish Identification Guides
                National Plan
                International Plan
                MCS Implementation Plan
                      data collection
                      special regulations and policies
                      enforcement strategy

Plan 3 -   Crustaceans, etc.
                Fish Identification Guides
                National Plan
                International Plan
                MCS Implementation Plan
                      data collection
                      special regulations and policies
                      enforcement strategy

Fisheries Habitat Management Plans

      Objectives
      Areas for special concern
      Monitoring and control procedures

Vessel Types and Markings

Fishing Gear Guidelines

MCS Operational Procedures
      Data Collection
      Boarding and Inspection
              non-hostile
              hostile/use of force guidelines
      Investigation
      Seizure
      Confiscation
      Disposal and security of goods seized
              perishables
              non-perishables

Prosecution
  evidence gathering and security of same
  detention and ordering to port for further investigation
  arrest procedures
  pre-trial actions
  trial procedures
  post trial activities
Patrol planning   air
                   land
                   sea
Pre-patrol briefing and check list
  safety equipment
  report forms
  inspection and data collection equipment
  gear check
  translation guide for inspections (if necessary)
Patrol report guides
Post patrol de-briefing check procedures
Communications guides
            radio frequencies for support stations
            radio procedures
            names and telephones of key persons for
              support, by Ministry

Reports and Documents

These include samples of the various reports with detailed instructions as to the proper completion of each document, where these instructions are necessary.

Common terms and conditions for licenses
Fisher's license
Vessel license
Vessel movement reports
Vessel inspection reports
Seizure reports/receipts/evidence tags

Fishing log interpretations for fishing, processing, freezing, transhipment and storage
Witness statements
Certificates, where these are appropriate
Data collection reports - biological, social and economic
Patrol reports

Port State Controls and Contacts

CORE COMPONENTS OF FISHERIES OFFICER TRAINING

Training requirements for Fisheries Officers will vary in every country due to many factors, not the least being the level of the fisheries knowledge and the corresponding standards of education in the State. There have been two major fisheries training needs assessments in the past few years, one in the South Pacific for FFA and a second in the Caribbean for the CARICOM Fisheries Resource Assessment and Management Program. Although the details varied with respect to the training required, it was noted that training should be an ongoing exercise and there should be a commitment to enhance the capabilities of fisheries staff as they increase their responsibilities. Training should correspond to the responsibility level of the position and be matched by the general educational level of the recipient. In the Caribbean, for example, the broad categories of fisheries field staff include fisheries assistants, fisheries officers, senior fisheries officers and chief fisheries officers/fisheries directors/fisheries administrators. Each of the general levels has several different titles. It was noted that the shortage of individuals with formal secondary education places pressure on those with said education to accept a much higher level of responsibility and, consequently, their own formal training and educational level needs to be at a high generalist/managerial level to give them the capability to address the wider range of issues, compared to countries where the human resource pool is sufficient to fund and attract specialists. This is more evident as the seniority of the individual increases.

Fisheries Assistants tend to be technical level officers who are in the infancy of their careers and consequently required introductory and hands-on technical training. It is expected that this level of officer would possess a minimum of high school or "A" level formal education to be able to respond appropriately to the training recommended. Officers below this level would be encouraged to upgrade their formal education. The fields of study for these officers should include, in random order, the following:

level 1 orientation	fishing methods
fish handling	navigation and safety
extension field work	communications
administration	resource management principles
basic biology	fishing gear design and construction
gas and diesel maintenance	vessel repairs and maintenance
small fish business practices	introduction to computers

Fisheries officers, on the other hand, are expected to possess a formal degree in a fisheries or marine related field of study. Their level of training for their duties emphasizes a higher theoretical level of resource management.

Their training should focus on the following:

level 2 orientation	resource assessment applications
aquaculture	data collection techniques
socio and economic principles	fish technology
fisheries development	fisheries law

marketing

sea use planning introduction

policy and planning skills

advanced computer skills

processing

extension field training

introduction to oceanography

Senior fisheries officers, with supervisors in the fisheries department, are expected to possess a minimum of a first degree. It was seen as advantageous if one has a post graduate degree in a related field. This level officer, as a supervisor and senior individual in the department, would be called upon to manage personnel and represent the department at meetings regarding ocean use management in general. The individual needs greater knowledge of the integrated ocean management principles and techniques to properly represent the department. The training suggested for this level of officer includes the following subjects:

level 3 orientation

environmental impact assessment

value added product skills

data interpretation and analyses

aquaculture

survey methodologies

human resource development

advanced fish marketing strategies

stock assessment

survey descriptions

finance and administration

project management

personnel management

general management skills

integrated sea use planning

Fisheries administrators are again expected to have a minimum of a first degree in a related field and considerable field experience. These individuals are the department's representatives to the government and require senior level management skills as well as knowledge and ability in the planning and policy side of fisheries and habitat management. Subject areas identified for this level officer include:

senior management orientation

policy development & mgt.

Convention on the Law of the Sea

international affairs

communications skills

personnel management

finance and administration

MCS strategy and policy development and implementation.[1]

program development & mgt.

fisheries dev. & mgt.

donor agency programs

socio and economic analyses

legal interpretation

senior management skills

This information is presented as one example of a regional initiative with respect to an assessment of fisheries training required to meet gradually increasing levels of responsibility in the fisheries departments. It might be noted that there appears in the above example a dearth of information on fisheries MCS, except at the final level. This may be true, but it could be expected that the general orientation and the fisheries management training would cover this area of responsibility. It might, however, be clearer if the MCS training commences at the first stage and progressively increases in profile and

---

[1] *O'Reilly, A. and Clarke, K. (1993).*

tasking at each level. Certainly it would be expected that officers would all require considerable training in this area of work with emphasis on the enforcement aspects of MCS, if fisheries are to carry out the surveillance aspects of fisheries management, or guide the other ministry officials seconded to them for this purpose. Fisheries administrators and their field staff would require the knowledge to guide these individuals appropriately in fisheries management techniques and priorities. It does appear common to all training packages for fisheries that the following general subjects are required for fisheries officials to carry out their duties:

resource management	habitat management
data collection and analysis	enforcement/surveillance
public relations	administration
personnel supervision	coastal zone planning and
and management	integrated management
policy development and implementation	

It has been found through experience that training for fisheries can best be achieved through the use of modular training techniques. This can assist in minimizing the time the officer needs to be away from duty and permits assimilation of the theoretical knowledge into practical experience between training sessions.

The emphasis and detail for each subject will be determined by the Fisheries Administrators to meet the individual needs of their countries. Training, especially on a regional or sub-regional basis, is an area of high interest to all donor agencies.

As this paper is focused on MCS, an expansion of that module might assist Fisheries Administrators in planning the training programmes for their staff. It must be noted that the fisheries officer will not be fully qualified to carry out MCS duties following this module, due to the fact that it is the linkage and knowledge of the other training modules which cements the capabilities into a competent whole.

Without setting priorities for the training, some of the task modules which might be concentrated upon during the training would include:

- understanding of the Department's mandate and jurisdiction
    * answer the questions as to why there is a fisheries department and
        what it is?
    * answer the question as to how far does its authority extend?

- the purpose and scope of MCS/the objective of the country's MCS policies
    * the purpose of MCS
    * difference between MCS and enforcement
    * difference between actual and preventive enforcement
    * departmental MCS policy
    * difference between renewable resource management and other
        management strategies

- principles of law
    * the purpose of the law
    * the role of society in establishing law
    * the impact of law on society
    * how laws are made in the country
    * the meaning of the law (interpretation of the law)
    * what is meant by case law, common law, civil law, summary conviction and
indictable offence
    * powers of search, arrest, entry and settling disputes as established in
        fisheries law

- the court system
    * how the judicial system works
    * levels of the courts and authorities
    * court terminology

- introduction to the fisheries laws - acts and regulations
    * how to interpret the Fisheries Act(s) and regulations
    * the lay-out of the Act(s) and regulations
    * the use of the Act(s) and regulations
    * authorities and powers of fisheries officials

- the support resources available to address the task
    * who controls the support services
    * how are these resources accessed
    * emergency support

- the co-operation and linkages with other ministries that are necessary for successful MCS
implementation
    * which ministries may become involved in fisheries MCS activities
    * what are their procedures which impact on fisheries MCS activities
    * who are the contact persons in these ministries
    * what is the official mechanism to interact with these officials
    * what is the mechanism in off-hours or emergency situations

- determining MCS priorities
    * identify MCS problems
    * identify problem area and the impact of continued activities
    * develop a plan and allocate resources

- planning MCS activities, data collection and surveillance patrols
    * routine patrol planning for land, sea and air patrols
    * dedicated patrol planning for land, sea and air patrols
    * measure fish, fish catches and fishing gear
    * collect scientific information through sampling techniques
    * collect socio and economic information through questionnaires
    * stop vessels at sea
    * stop motor vehicles

        * develop local contacts and sources of information/community relations
        * conduct checks of licenses,vessels, gear, vehicles, facilities and
            persons

- determine the violation
        * evaluate the situation
        * conduct searches
        * evaluate complaints

- apprehend violators
        * establish identification of self and alleged violator
        * advise alleged violator of offence
        * obtain information from the alleged violator and witnesses
            warnings
            note-taking
            exhibits, care and security
            interview techniques
            use of recording equipment
            questions to ask
            evidence, definition and use thereof
            definition of threat, promise
            elements of a charge
            separation of alleged violators
        * issue appearance notices, warnings or secure certificates

- arrest alleged violators
        * ensure custody
        * searches
        * rights of the alleged violator
        * release from custody
        * appearance notices
        * policy on use of force
            issuance of firearms
            policy on use of firearms
            practical firearms training
            armed boarding training
        * application of use of force
            procedures for escalation of use of force

- gathering evidence
        * maintain the scene of the alleged violation
        * make and secure seizures
        * obtain exhibits
        * continuity of evidence
        * statements
        * liaise with senior staff and counsel

- charge alleged violators

    * violation reports
    * prepare information
    * summons
    * laying of information and summons
    * serving summons
    * proof of service of summons
    * subpoenas
    * executing warrant of arrest
    * orders of forfeiture

- court procedures
    * court duties
    * giving evidence

- completion of final documentation
    * return of prosecutions
    * court case follow-up action

- completion of final procedures
    * return seized items or proceeds
    * disposal of forfeited items

-review and evaluation of MCS program
    * assess MCS activities
    * recommend amendments to procedures or control mechanisms
    * recommend enhancements to MCS procedures, equipment, staffing, training, etc.[12]

---

[12] *The CFRAMP Training Needs Assessment, FFA Training Needs Assessment, training programs in Canada, FFA, Belize and USA have been used extensively to produce this summary of training modules.*

CORE COMPONENTS OF OBSERVER TRAINING

This Annex is very dependent upon the objective of an observer program. If it is assumed that observers will be employed by the State, or a third party acting on behalf of the State, and they will be used for data collection and advice to the vessel master only, **with no enforcement powers**, then the following could comprise elements for their training. The material for this section has drawn heavily on the FFA and Canadian experience. These have been relatively good experiences, but it has been found that without the Government commitment to take the observer programme seriously, it can be a very abused fisheries management tool.

Problems which can be encountered include hiring practices, poor work practices, lack of commitment, lack of funding received at the programme level and lack of basic educational skills. The personal integrity of the observers is an important factor, as fabricated data sets, if used can distort the fishery management advice and hence impact very negatively on the fish stock assessments. These points are brought out to note the requirement for full government commitment and support for this programme, if it is to succeed. This is not a programme that is appropriate for all countries, and where this is the case, it should not be used.

The complexity of the observer programme can be assessed and decided by the Fisheries Administrator in accordance with the needs of the State and the level of competence of the staff available. As with fisheries management, each observer programme should be custom designed for the State. The following, therefore, is a listing of elements and some detail on the lecture content for observer training which can be drawn from as required by the Fisheries Administrator in the design of the programme.

COMPONENTS OF AN OBSERVER TRAINING COURSE

1. Role of Observer

The role of the observer will be stressed in that he may only observe, record and report. The methods by which each of the aforementioned is accomplished will be addressed. It will be emphasized that part of the reporting requirements is to advise the master of irregularities.

The appropriate regulations concerning observer safety and rights will be addressed for both foreign and domestic vessels. The intent of these regulations will be explained.

In an introductory lecture on the duties of an observer, the two principle aspects of the job should be emphasized.

1) Monitoring compliance of fishery laws,
2) Conduct biological sampling.

The fact that each of these basic principles is complementary to the other will be explained. The basic tools: observing, recording and reporting will be emphasized.

2. Introduction to Fisheries to be Observed

A brief lecture on the fisheries to be covered by fisheries observers will be given noting the fish species and common gear used to catch these fishes. This will be broken down into national and international fishing authorized in the zone.

3. Introduction to Fisheries Laws

These sessions are pertinent if the observer is hired to monitor compliance of the vessel and crew with the national fisheries legislation. The sessions would be structured to cover, in detail, the various acts and regulations for fisheries with particular emphasis on areas of concern for fisheries management. Some of the subjects would include:

Licensing
Authorized fishing areas
Authorized fishing gear
Fisheries management plans for each species
Records of fishing
Prohibitioned catches
Unauthorized fishing activities

These sessions will also address the problem areas in the fisheries with respect to compliance with fisheries laws. These could include misreporting in the logbooks through mis-representation of the conversion of fish from product wieght, or fish reduction, back to whole, round weights; area violations; double book-keeping of catches, one for the inspector and the real one for master; discarding and dumping.

4. Catch Estimation

This session is to note the various catch estimation procedures available to the fisheries official/observer to verify the actual catch of the vessel. It will also note the dificulties of estimating when mixed species are involved.

1) a. A brief introduction as to the importance of estimating catches as accurately as possible, explaining that the estimate of total catch is the most reliable estimate and that species in the least quantities should be subtracted from the total catch with the remainder assigned to the species in the greatest abundance.

b. Emphasize the importance of accurate estimates in relation to monitoring the vessel's catch recording/reporting practices.

2) Total Catch Estimation techniques are discussed

   a. Codend estimation once on deck, highlighting the use of strengthening straps.

   b. Utilizing the bunker which holds a known quantity which can be determined by interview and/or

   c. Use of baskets to determine a density figure applied with a measurement of the bunker.

   d. Crew member estimates
      i) Caution on estimating of catch in processed weight,
      ii) Caution on the possibility of misleading information,
      iii) Comparison of icer's figures to logbook's figures in relation to determining whether dressed or round weights recorded.

3) Species Composition Estimates

   a. Utilization of percentage estimate while catch is being dumped.

   b. An actual estimate of weight in codend when species are of small quantities during dumping of codend.

   c. An estimate based on how many baskets that species would fill, while being dumped, compared to what a basket of that species would weigh.

   d. Observation of the catch being processed - know what percentage of catch has been processed and compare it to the percent estimate of species observed to have been processed and extrapolate back to total catch.

   e. Observation of catch culled in factory.

4) Verification Techniques

   a. Bunker capacity - density
   b. Crewmember estimates
   c. Icer's figures/processed fish

5) Discard/Reduction Estimates

    a. Weighing of all fish

    b. Weight of fish/time period/processing time

5. Conversion Factors

1) The definition of a conversion factor as associated with the fishing industry will be explained. The source of conversion factors to be used will be discussed.

2) Symbols such as pie's will be applied using the percent (100%) concept in order that individuals may understand how a conversion factor is derived.

3) Given a known quantity that can be converted back to a whole (100%) by utilizing a conversion factor, what percentage would be non-utilized material.

4) The concept of the already compensated for material which was not utilized, being produced as a by product thereby not requiring conversion will be put forward. The pie concept will be used.

5) Finally this will all be drawn together by substituting the pie for a fish undergoing processing.

6. Gear Type

This training is focused on the identification of various fishing gear, their component parts and how to measure the parts to ensure compliance with fisheries laws, if there are such, pertaining to mesh size.

1) A brief discussion on the importance of being able to identify gear types and component parts to ensure compliance with the authority in the fishing licence.

2) Examples of component parts of trawls will be shown by utilizing diagrams and a model with a brief explanation of the purpose of each component given. Modifications will also be discussed at this time.

3) Diagrams of various gear types will be shown highlighting the differences between gear types. Distinctions will be discussed.

4) Utilizing a diagram and a model, indicate what measurements are necessary to ensure the fishing gear is measured in a fashion acceptable to the courts of the land.

7. Introduction to Navigation

Part of every observer's training is to know where the vessel is fishing. This mini-session on navigation will assist the observer in this regard.

1)   Latitude and longitude will be explained as to the orientation on a chart. The component parts will be shown: degrees, minutes and seconds, explaining the significance.

2)   Basic plotting of a position will be shown with each individual having to plot several positions.

3)   Distance travelled between points in relation to speed and time will be discussed.

8. Biological Sampling Methods

This would be a lecture on biological sampling methods highlighting proper sample collection.

    a)   Random samples
    b)   Stratified samples
    c)   Combined samples
    d)   Processed samples
    e)   Discard samples
    f)   Reduction samples

9. Species Identification

Every observer needs to know the identity of the species which are being fished in the country's waters. There needs to be at least one session to ensure the observers are very familiar in this process.

10. Sampling Techniques and Requirements

Where it is decided that observers will also be utilized to take biological samples for the resource assessment activities of the stock assessment personnel, the individual will need to be trained in various sampling techniques and standards in accordance with acceptable scientific procedures. Some of these are noted here for reference.

1) The actual types of samples taken in relation to their importance in the overall sampling scheme of the State will be noted.

    a)   directed species
    b)   bycatch species
    c)   reduction
    d)   discard

e)   otoliths

f)   specific requirements

2)   Numbers of individuals comprising a sample will be addressed.

3)   Sample selection will be reiterated.

4)   Length frequencies will be discussed with demonstrations.

5)   An actual length frequency will be done by each candidate.

6)   Sexing of fish will be demonstrated. Each candidate will obtain hands on experience.

7)   Otolith collection will be demonstrated on various species. Hands on experience will be obtained.

8)   Morphological requirements will be discussed and demonstrated. Hands on experience will be obtained.

9)   It will be emphasized that full morphologies are required while taking otoliths and it will be stressed that during morphologies all information with the exception of collecting the otoliths is required.

11. Logbooks

The fishing records of the activities in the country's fisheries waters are the only real written record of events. Observers must be familiar with all aspects of fishing records to ensure they are being completed correctly. This includes ensuring all the data are being recorded regularly and accurately. Such data might be:

1.   Date, licence number, activity, position, time, depth, gear, mesh size, retained and discard estimates by species.

2.   Format for fishing sets by gear and the correct times.

3.   Licence number, vessel name, side number, date and the proper entry of species by product form.

4.   The proper determination of meal produced from offal, considering product produced, will be explained. The lecture will consider individual species and product forms produced. The approach used will be:

a)   Identify what products have yielded waste that will go to meal. Emphasize products frozen round do not yield offal.

b)   Has any of the waste material been utilized to produce a by product. Identification of some of these products.

c)   Convert all product forms to round that have produced waste that will go to meal.

d)   From the total round weight determined in step (c) subtract the product weights of all products that have been converted to round. Also subtract the weight of any material identified in step (b) as having been produced as a by product.

e)   This is now the total round weight of waste material that will go to meal. This figure divided by the fishmeal conversion factor provided by the vessel will give the amount of meal (product wt.) that will be produced.

5.   Determination of round fish produced as meal will also be discussed.

a)   It will be noted that estimates of round fish will often be utilized by the vessel to arrive at this figure.

b)   The utilization of appropriate conversion factor for meal production will be explained. The product weight can then be converted back to round wt.

c)   The fact that excess meal production must be reported as round fish to meal will be discussed.

6.   Problem Areas - Areas where misreporting can occur will be discussed. Specifically the recording of estimates for kept and bycatch species, area of capture and start position will be covered.

12. Documentation of Irregularities

The three places in which information pertaining to infractions are documented, and the purpose of each, will be discussed:

1)   Notebook - purposes:
     a) to aid in the monitoring process;
     b) to aid in report writing;
     c) to aid in accurate testimony;

Ongoing entries should be made at the first possible instant.

2)   Observer Diary - purpose is to provide a detailed chronological documentation of trip events, used to assess situation and decide whether or not to proceed with charges. Enteries are made at regular intervals (i.e. each evening).

3)   Observer Trip Report - purpose is to provide a concise reference to the irregularities. Entries are made throughout the trip and at trip end.

Documentation Rules for the notebook will be addressed and explained. The contents of the documentation of an irregularity will also be discussed (Answering the questions who, what, where, why, and how, along with extraneous information such as weather/Personal information or opinions should not be recorded).

13. Observer Trip Report

The most important document from the observer is the trip report and as such it should be as complete as possible. The following are some of the areas which Fisheries Administrators may wish to include in the requirements for their observers in their final trip reports.

Vessel Information	Observer Activity
Daily Trip Summary	Comparison of Observer Estimates & Vessel's Fishing Log
Sampling Inventory	Vessel Sighting
Fishing Pattern	Unique Areas
Fishing Operations	Logbooks
Discards	Hold capacity
Trip Summary	

Please note that this is only one example of possible observer training scenarios and all modules are not essential, nor ideal to address the fisheries management situation in each case. Each Fisheries Administrator can pick and choose various modules appropriate to the country's fisheries and adapt them as required.

CORE COMPONENTS OF MCS REPORTS

This annex is intended to provide examples of the elements included in existing report forms for consideration by Fisheries Administrators. It is recognized and suggested that the uniqueness of each MCS system will necessitate that each Fisheries Administrator will wish to design report forms to meet their State's requirements. It is for this reason that actual report forms from countries, which would soon be outdated, are not presented, but instead the core information to be included in these reports is suggested in lieu of the former.

1. LICENSE APPLICATION

The following information is common to collect for license applications. This is the first document which will set up the information database, consequently the information collected here is crucial for accuracy in identifying the vessel.

name of vessel,
country and port of registry,
registered number,
international radio call sign (for vessel marking and identification),
side number (if different from the radio call sign),
type and class of vessel (longliner/stern trawler, etc.),
length overall,
registered net and gross tonnage,
engine type and power,
description of the vessel (construction material, year built, colours and profile, sometimes a picture is requested),
fishing gear aboard,
communications equipment aboard and listening frequencies,
name and address of owner with fax number and telephone number,
name and address of vessel master,
name of the representative for the vessel in the country,
number of crew,
hold capacity and type (wet freezer),
processing equipment,
freezing equipment.

The application would also include the request for the fishing privilege in accordance with the State's requirements, the fishing plan.

2. VESSEL MOVEMENT REPORTS

a. Zone Entry and Exit

date/time of report,
vessel name,
vessel call sign,
vessel side number (if diferent from the call sign),
date of entry into/exit from the EEZ/fisheries waters,
position of entry,
weight of fish onboard by species and product form
intended area of fishing (This is after the first entry. First entry should result in a visit
to the regulatory port for a briefing.)

b. Port Entry/Exit

date/time of report,
vessel name,
vessel call sign,
vessel side number (if diferent from the call sign),
estimated time of arrival/departure (ETA/ETD) to port
designated port

c. Area Change for Fishing

date/time of report,
vessel name,
vessel call sign,
vessel side number (if diferent from the call sign),
current position,
area for intended fishing
time of entry into area

3. CATCH AND EFFORT REPORTS

These would be in a format **and time frame** as set by the coastal State.

date/time of report,
vessel name,
vessel call sign,
vessel side number (if diferent from the call sign),
current position,

Fishing report - most countries require the vessel master to provide data on the position
at a standard time each day and a summary of catches for the period from the same
time the previous day.

date,
time,
number of sets,
number of hooks/type of gear,
total fishing time that day,
catches by species,
total daily catch.

This report is sent to the fisheries authorities as required. Some countries require this each week and others, daily.

4. LOGS

Logbooks pertaining to fishing operations are as varied as the number of countries and companies fishing. It is for this reason and for ease in computerized data entry that some countries issue their own logbooks for all vessels fishing in their waters. The information collected usually falls into three main categories, fishing, processing and transhipment.

a. Fishing

Fishing logs commonly require information similar to the catch and efort report, but in a more detailed fashion:

vessel name,
side number,
license number,
date,
position at the set reporting time,
area being fished,
target species,
time commenced for each set or tow,
time of completion of each set or tow,
hours fished,
position at the start/end of each set or tow,
type of gear,
number of hooks/lines/nets,
depth of fishing where applicable,
catch by set or tow by species and weight/size,
discards,
round weight processed for human consumption,
round weight of fish reduced to meal,
cumulative totals,
surface sea temperature,
observations - sea, currents, weather, wind, temperature, etc.
activities other than fishing/remarks.

b. Processing

vessel name,
side number,
license number,
date,
product form by species and weight, (frozen round, gutted, gutted head on, fillets, salted, pickled,canned, oil, etc.)
meal,
cumulative totals,
remarks.

c. Transhipment

sending vessel name,
side number,
license number,
receiving vessel name,
side number,
license number,
position of transfer,
date and time of transfer commencement/completion,
product transferred by species, product form and weight,
cumulative totals,
remarks.

This information can be cross checked against the catch and effort reports, observer reports, position reports and sightings to verify the accuracy of the reports. This information can be utilized for patrol planning as well as for the biological assessment of fish stocks. It is recognized that all information is not required for all fishers, but the majority of this information from large vessels can be of assistance to fisheries management and planning, including MCS operations.

5. VESSEL SIGHTING REPORTS

These reports are fairly standard from both sea and air sightings. The main components include:

vessel name,
side number,
nationality,
description of the vessel,
vessel type,
position,
activity (course, speed, fishing, etc.),
licensed/unlicensed

6. VESSEL INSPECTION REPORTS

These are the reports which are used to collect additional data on the fishing operations of vessels and also for the verification of the reports sent by the vessel to the fisheries departments. These at-sea and in port inspections, when conducted carefully, will assist the Fisheries Administrator in confirming the vessel master's compliance with the country's fisheries laws. The following are the common generic components of fishing vessel inspection reports:

vessel name,
port of registry,
nationality,
vessel type,
length/breadth/draught,
gross registered tonnage,
net registered tonnage,
fish processing capacity,
fish storage capacity,
fish processing equipment,
freezers/capacity/frozen storage,
side number,

license number,
validity for fishing/area/species/dates,
date and time of inspection commencement and departure from the vessel,
name and address of the master,
name and address of the owner,
name of the inspector,
name of vessel carrying the inspector,
position as determined by the vessel master,
position fixing equipment,
position as determined by the inspection vessel master,
position fixing equipment,
fishing gear on deck/type/material/attachments/net measurements/number of hooks etc.
number of crew,
estimate of fish caught since last inspection by species and weight/product form etc.,
estimate of fish on board,
transhipments of fish/to whom/species/weight/when/where,
fish processed since last inspection,
discards,
fish meal/oil produced,
summary of fishing from logs/species/area/weight/dates,
records inspected,

last port of call/dates,
next port of call/dates,
apparent infringements,
photographs taken,
comments from the inspector,
comments from the vessel master,
signatures/dates,
witness signatures/dates,
copy of report left with the vessel master.

CORE COMPONENTS OF A FISHERIES MANAGEMENT PLAN

Each fishery, or group of fisheries, that is being harvested should be done in such a manner as to ensure the conservation of the species. This is done through management of the resource. There are several tools for this purpose, some that concentrate on direct management and others that focus on indirect management of the resource through management of the harvesters for various reasons. Many countries mix-and-match the two in an effort to manage the resource. This, on occasion, can result in an intrusive government policy which may be perceived as very restrictive to the fishers. Each management plan should have an objective, principles or assumptions and a clear concise plan that is able to be implemented effectively at minimal cost. Above all, each management plan must have the acceptance of the fishers to whom it applies or it will be doomed to fail.

An example of a framework of a management plan for large pelagics follows:

Objective:

Conservation and restoration of the fisheries resources for the economic benefit of the citizens of the country.

Principles:

1.   Allocation of resources will be on the basis of the terms of the Convention on the Law of the Sea, that is to say, the national fishery will take priority and excess resources may be allocated to interested international fishing partners.

2.   The allocation of resources to the national fishing sector shall be based on the historic dependency and catches of the various fleet sectors considering also the mobility of each sector.

3.   The artisanal fleet sector shall receive priority access and area protection due to the lack of mobility of the sector and the dependency of the coast communities on the fisheries. In return, the community fishers will be expected to assist in the MCS activities in their areas.

4.   All mobile fishing gear fishers shall meet to agree on the division of the excess resources not required by the artisanal fishing sector, the latter defined as that sector which does not normally stay at sea more than one day due to the size of the vessel and fishing equipment. The division of the fishery resources between the mobile sectors shall be monitored through effort surveillance and results passed to the sector for their control purposes. The state shall close the fishery when the effort level for the fishery is expected to be reached. There will be no new entrants to the mobile fishery.

Note: The principle here is that the fishing sector must police any division of the resource themselves and the State will concentrate on conservation of the total resource. This can be expected to create difficulties initially, but the State should remain focused on conservation of the total resource, not the parts. As always the exception to the rule as seen here is the coastal fishery which becomes very vulnerable to larger mobile fishing gears.

5. All vessels and fishing gear shall be inspected prior to the issuance of an annual license to each of the fishers and the vessel. Only authorized fishing gear can be carried on the vessel during fishing operations.

6. Access by international fishing partners shall be subject to principles under the Convention on the Law of the Sea. The fisheries agreement and license fees shall take into consideration the value of the resource to be harvested, contribution to the scientific data bank for fisheries management, access to all records for the fishing vessels and their activities both inside the fishing zone and the adjacent high seas, access to final weigh outs of the final product form, historic compliance of the fleet with fisheries laws of the State, or other states both inside fishing zones and on the high seas, training and employment opportunities for nationals, financial contribution towards recovery of costs of conservation of the resources and contribution to the national economy.

7. International fishing partners shall carry observers, for data collection and advice to the vessel master for their allocated fisheries under the conditions set by the state and reimburse the state for the costs of these individuals.

8. Transhipment of fish for international partners shall be in one of the named ports for which the port authorities shall assist in facilitating the exercise in a timely and cost effective manner.

9. The allocation of fish and the establishment of fishing fees shall be on effort in the fishing zone.

10. The fishery may be closed, or extended, by the state at its prerogative.

Management Plan: (EXAMPLE ONLY - all names are fictitious)

Artisanal Fishery:

The coastal zone for artisanal fishers extends to 15 kilometres from the baselines around the coast and islands. No mobile fishing from vessels over 12 meters shall be authorized in this zone.

The coastal zones for artisanal fishers shall be divided into four areas, north, east south and west. The area definitions shall be between the following points:

North - from a true bearing of north from the church on Samuel's Head east to a bearing of 090 degrees true from the quay in Domingo Bay.
    Effort allocation - 20,000 days

East - from a true bearing of 090 degrees off the quay in Domingo Bay east and south to a true bearing of 180 degrees off the northern entrance to Hungry Tickle.
    Effort allocation - 42,000 days

South - from a true bearing of 180 degrees off the entrance to Hungry Tickle south and west to a bearing of 270 degrees true off Fishers Light.
    Effort allocation - 42,000 days

West - from the bearing noted off Fishers Light to the bearing of true north on Samuel's Head.
    Effort allocation - 32,000 days

The communities noted in the boundary descriptions have agreed to host the fish zone committees for fishers in each zone. A federal fisheries representative shall assist with these meetings on fisheries affairs. The committees have the authority to issue licenses and set fishing fees under general fisheries guidelines. Federal fisheries shall monitor fishing effort and advise the committees of the effort expended on a monthly basis until the full effort allocation has been utilized. Community elders are requested to assist in the data collection of effort expended and reporting of non-compliance with closed seasons.

Mobile Fishery:

The mobile fishery zone shall extend from 15 km from the baselines as noted in the attached chart (chart to be attached) to the edge of the EEZ.

The total effort for the large pelagic fishery for this zone is 200,000 days.

Representatives from the seven major fishing communities involved in this fishery have agreed that the shares for the fleets from each community shall be pooled together and monitored as one fishery for the 900 vessels.

Fishing shall be permitted all year.

Vessels shall be authorized to carry and use only the fishing gear for which they have been licensed.

Federal representatives will report monthly effort figures to the representatives of each of the major fishing communities for this sector and advise fishers when the allocation is nearing depletion.

International Fishing Partners:

The total fishing effort allocation for the international partners is 20,000 days.

Currently only the vessels from "Partner's Country" has reached an agreement on fishing in the zone. The allocation for this fleet of 20 vessels is 3,000 days.

This privilege has been granted on the basis of licensing and fishing fees totalling $4,000,000 U.S. Additional fees for observers for all vessels total $600,000 U.S. A further $2,000,000 U.S. has been granted for scientific research assistance in the country of which 35% is for national salaries. Each vessel shall carry four national fishers for on-job training in the offshore fishing fleet. The weigh outs of fish caught in the zone shall be forwarded within thirty days of landing.

Each fishing vessel shall enter the port of "Home Port" to pick up its license, observer and a meeting with fisheries officials. The vessel shall be inspected at that time and only fishing gear authorized in the license may be carried on board the vessel. The reporting requirements shall be contained in the fisheries guidelines provided to each master.

Vessels may only fish in a zone outside the artisanal zone shown on the chart provided.

No vessel shall set a fish aggregating device (FAD) within five nautical miles of the government set FADs. The location of the latter will be provided to each master in the fishing guidelines.

Vessel masters shall assist observers in the execution of their duties in accordance with the regulations. Vessel masters shall adhere to all directions provided by fisheries officers in the execution of their duties and the full assistance of the vessel master and crew is expected.

This may perhaps be an over-simplification of the process, especially when mixed species fisheries may be carried out in the same zone, but it is provided as an example of a possible scenario. Where mixed fisheries occur, the effort control can be averaged for the fisheries and allocated accordingly with a measure of caution inserted on the side of conservation. The back up is the fact that actual landing figures can be obtained for the artisanal fishery and the national mobile fishery while observer coverage can assist in the international figures. If there is a concentrated effort towards one of the species due to its economic value, the fishery can be closed. These options can be discussed fully during negotiations and talks with the fishers and international partners. The mixed fisheries should be established on the basis that conservation is still the priority of the government and should be the priority for the fishers as well. The final control of the fishery remains in the hands of the government and it should always be exercised on the basis on conservation.

<u>COMPONENTS OF THE PROSECUTION PROCESS</u>

This annex is compiled mainly of pertinent excerpts from the prosecution manual for the South Pacific Forum Fisheries Agency. In reviewing the various initiatives for common law countries on the subject, the FFA manual appeared to be the most advanced and comprehensive and key sections are hereby presented (in total) for information and reference. The manual was funded by the International Centre for Ocean Development (ICOD) and written by Mr. R. Coventry.

The present annex has been included in this paper as it is seen as being of considerable practical importance. However, a number of important qualifications have to be made regarding its inclusion in the present publication. First, the points made concerning prosecutions in this annex are made very much against a common law background. Many of these points would arise in very different circumstances in civil law systems. In fact, a very different approach to that contained in this annex would be necessary in order to do justice to the issues that might arise in a civil law system in respect of a prosecution in such a system.[1]

Second, even within the context of the common law, the present document reflects concerns of prosecutors in Australia and certain Pacific Island States, which are to some extent influenced by local practices, particular provisions of the law of evidence not necessarily applying in other jurisdictions, possible restraints flowing from the constitution, and the relevance of particular judicial decisions.

Third, the special focus of the present annex being on **prosecutions**, it should be stressed that other factors might need to be considered, namely, whether there are other avenues to pursue in individual instances, such as compounding or administrative penalties, or even whether in some situations, diplomatic negotiations would be prefereble.

Accordingly, the present annex is included here with an important caution that, despite its intrinsic value in alerting Fisheries Administrators to a number of matters relevant to prosecutions against fishers, it has to be used with care in those parts of teh world for which it was not designed. Furthermore, the practices on such matters are apt to evolve within individual jurisdictions, and these merit constant evaluation.

---

[1] *It was beyond the competence of the author to expound on case scenarios in this latter situation, consequently, the development of guidelines for civil law countries for fisheries will, of necessity, be the subject of further consideration and future revisions to this publication.*

PREPARATION FOR TRIAL

This section is divided into two parts:

Part I   General Preparation

Part II   Pre-Trial Checklist

Careful and detailed preparation of the case is essential.

PART I GENERAL PREPARATION

A. EVIDENCE

It is for the prosecutor

- to prove his/her case
- to prove it beyond reasonable doubt
- to prove it for each charge and each defendant
- to prove it by admissible evidence.

1. Witnesses

The original copies of all witness statements, in the correct form and signed, will be needed.

All the statements must be read, and the following checked:

(a)   Do they prove the charges that have been/will be laid?

(b)   Do they rebut any defence raised in conversation with the defendants or formal interviews, (e.g. the master alleges engine breakdown - witness statement from engineer to disprove this)?

(c)   Do they contain inadmissible or unfairly prejudicial statements, (e.g. hearsay, references to criminal matters not charged)?

(d)   Does the witness refer to the exhibits he is producing?

(e)   Is an additional statement required from the witness to clarify anything or add anything useful?

(f)   Do they aver the reasonable suspicion, reasonable belief or other basis which must exist before the witness can act, (e.g. for boarding/apprehending/hot pursuit)?

(g)   If statutory presumptions are to be utilised, do the statements lay the requisite foundation of fact, (e.g. some Fisheries Acts provide that if an officer suspects any fish to which the charge relates were taken in a particular area of waters and he gives evidence of the grounds on which he so suspects and the court thinks the suspicion reasonable then in the absence of proof to the contrary the fish will be deemed to have been so taken).

Every fishing case will be different and, of necessity, the evidence will vary from case to case. Only experience and background reading will fully equip an advocate to appreciate all the evidence that can be utilised in a fisheries case.

2.   Useful Observations

Here is a list of the kinds of observations which might be used as evidence of recent fishing.

(a) On Sighting Vessel

(i)   Hasty departure - sudden increase in speed, clouds of engine exhaust, bow wave, anchor being quickly hauled in.

(ii)   In the water - dead fish, offal, seabirds picking at objects, muddy water drawn up from bottom (shallow water), sharks taking rubbish.

(iii)   Gear - buoys, flags nearby, wires, ropes dangling over side, small boats at reef, by vessel, being hauled in, divers.

(b) On Boarding Vessel

(i)   Bloody water and offal running from scuppers.

(ii) Noise of engine being started,[may or may not be pertinent depending on the fishery].

(iii) Crew - hastily stowing gear, bringing in anchor, pushing objects out of sight, generally agitated, wet or look as though been diving, fresh cuts and scratches.

(iv) Decks - wet running blood, offal.

(v) Gear - wet, and/or not stored/secured, wires still attached to gear, winches not disconnected, diving gear lying about or in small boats.

(vi) Fish - fresh, lying about deck or elsewhere.

(c) On Inspection of Vessel

(i) Wheelhouse - entries in logs, marks on charts, dummy logs, charts. Instruments working? Radar setting, satnav reading, echo sounder etc.

(ii) Freezers - fresh, half frozen fish, colour of eye, gill. Temperature temporarily higher than usual, temperature records showing rises and falls, machinery working, no signs of breakdown/repairs.

(iii)Engine room - main/auxiliary engine.- hot or cold. Test-run, no signs of malfunction or recent repair. Refrigeration machinery working. Engine temperatures, log variations.

(d) On Passage and Entry to Port

(i) Engine functioned well on passage to port.

(ii) Freezers operated normally.

(iii) Master had no difficulty in navigating and instruments worked.

(iv) Gear tidied up by crew on passage (c.f. photograph of state of gear when first boarded).

3. Exhibits

Ensure that all exhibits have in fact been collected. (It is very embarrassing to ask a court for an adjournment so an exhibit can be found).

The exhibits must be clearly labelled and protected and preserved in the most appropriate manner.

Examine the exhibits. Extra evidence can often be found, e.g.:-

Charts	rubbed out lines at area in question; EEZ, and closed areas faintly marked on;
Freezer Logs	fluctuations of freezer temperatures at times of alleged fishing;
Logs	twelve hourly positions not consistent with master's version of events; distances allegedly run impossible in time stated; is log a dummy? are entries consistent with other logs and charts? and
Photographs	show different stowage of gear on boarding from when vessel arrived at port.

Some exhibits require special care or procedures, e.g.:-

Photographs	the person who took them must produce them; has he retained the un-retouched negatives?
Fish	do they need to be kept as exhibits; can they be kept as exhibits? note any statutory power to sell and retain proceeds of sale, or dispose of if unsaleable.
Vessel itself	has it been immobilised; is there a power to immobolize? what is the emergency procedure for e.g. heavy storm, cyclone?
Radio Buoys	these can be damaged if dropped or knocked; they are expensive.

When exhibits are returned to defendants after a case they often claim that some articles have been damaged or deteriorated while in possession of the authorities. Care must be taken to ensure this does not happen and that false claims won't succeed. If a vessel is detained regular checks must be made, particularly of the freezers, freezing machinery, fish, engines and gear. Records must be kept of these checks.

4. Certificates

Many fisheries acts provide for proof of certain facts by certificate. These provisions should be used to the full:-

Check:-

(a) what facts can be proved by certificate;
(b) who issues the certificate;
(c) the form it should take; and
(d) its evidential value (conclusive proof, rebuttable presumption).

5. Machine Evidence

When reading the witness statements a note should be made of any evidence which relies upon scientific instruments, (e.g. readings of a satellite navigation machine, radar).

There is a common law presumption that the readings of notorious scientific instruments are accurate. An instrument will fall into this class if by general experience it is known to be trustworthy and so notorious that no evidence is required to prove it is trustworthy.

If a scientific instrument is "notoriously reliable", then readings from it can be given in evidence once it has been established that it was operating properly and the witness was a competent operator. However, the law is slow to recognize new instruments. For example, radar readings will be admissible whereas those of a satellite navigation machine will probably not be. The readings of an instrument which is not recognised as being "notoriously reliable" can still be made admissible if:

a) the whole system is proved by expert witnesses (a long and expensive process); or

b) the instrument was cross-checked against accepted instruments and found to be working properly and accurately.

Therefore, if a scientific instrument has been used, check:-

a) is the evidence needed?
b) is the instrument recognised by the court as "notoriously reliable"?
c) if not, was it cross-checked with instruments that are so recognised, before and after the material events? and
d) was the witness a competent operator?

It is also important to ascertain the maximum possible error of the instrument when functioning properly, (a satellite navigation machine will be accurate to a fairly high degree immediately after a satellite pass, but becomes progressively less accurate as it "dead reckons" positions until the next pass).

If the readings of an instrument are to be used in evidence then the prosecutor should go and see one and acquaint himself with its workings before the trial commences.

The most effective way of avoiding these problems is to have a statutory provision deeming the readings of instruments prescribed by the Minister admissible and prima facie proof of the facts averred.

6. Experts

An expert is a person with special skill, technical knowledge or professional qualifications whose opinion on any matter within his cognisance is admitted in evidence, contrary to the general rule that mere opinions are irrelevant. It is for the court to decide whether a witness is so qualified as to be considered an expert.

Experts are expensive and don't always form opinions which help the case of those engaging them. However, any case involving technical matters will at least require the assistance of an expert and possibly his evidence.

The first question to ask, therefore, is "Might the advice of an expert assist the prosecution case?" If the answer is "yes" then make sure the correct one is engaged - it is not a good idea to send a navigation expert into a freezer. He/she should also be well briefed as to which matters his/her opinion will be needed upon, although general comments will also be useful.

Always make an estimate of the likely cost before engaging him/her.

The decision to engage an expert might be made at a number of different times. Listed below are some of the types of expert that might be involved in a fishing case, when they might be engaged and for what purpose.

a) Expert on Fishing Methods - to inspect the seized vessel immediately upon arrival at port; to report on the type and the state of gear; to say how it is normally stowed; to detail evidence of recent use; to explain the general method of use of gear to the prosecutor; to state if the gear on inspection in port is placed differently to that which appears in photographs taken or observations made immediately after boarding; and to explain to a court in detail the method of fishing involved;

b) <u>Expert on Fish</u> - to inspect the seized vessel as soon as possible and comment upon the state, in particular the freshness, of all fish on board; to identify the species of fish aboard, and total weights; to give opinions on the factors affecting and rate of spoilage of fish in different circumstances;

c) <u>Engineer</u> - to inspect and report on the vessel's engines; in particular to report if there is any evidence of a recent breakdown or repairs to the vessel's engines, winches, etc;

d) <u>Freezer Engineer</u> - to inspect and report on the freezing gear; in particular to report if there is any evidence of a recent breakdown or repairs;

e) <u>Navigation Expert</u> - to give evidence upon how the position of the apprehended vessel was fixed (if he was aboard the patrol vessel); to relate the readings of machines involved, what checks were made thereof, how they work and possible percentage error;

f) <u>Valuation Experts</u> - to give advice and if necessary evidence upon the value of a seized vessel for bonding purposes, (this would avoid concerns which could arise when a vessel has been released on a bond of $100,000 if on "forfeiture" its real value turns out to be $500,000); and the value of the fish aboard.

Do not be deterred if an expert with the requisite list of qualifications does not exist in your area. Courts will accept the opinion of a witness who shows he has long experience of the matter concerned and is giving a considered opinion. "I've been fixing those engines and ones like them for years and there was nothing wrong with that one!" is perfectly good expert evidence.

Expert evidence can often be used to rebut defence put forward by vessel masters and other potential defendants, (see Common Excuses section).

Ensure that the witness statement of an expert commences with a statement of his/her qualifications and experience.

7. Explanatory Pictures

It is probable that the judge or magistrate trying a fishing case will little or no knowledge of the fishing industry or anything associated with it. All technical matters must therefore be carefully and clearly explained in opening and evidence adduced later to prove them. It is, therefore, worthwhile in such cases to prepare large explanatory diagrams and pictures. These can be used in opening and during the trial to assist witnesses in their descriptions.

e.g., the purse seine method of fishing:- a picture of a purse seiner with the principal parts labelled; four or five diagrams showing the various stages of a set with the operative parts labelled.

B. **POWERS OF OFFICERS**

The Fisheries Act and supporting regulations will set out who are authorised officers and what their powers are. Ensure that the documents appointing the authorised officers in the case are available, in the proper form and apply to the legislation concerned. It is common to automatically make police officers and masters of government vessels authorised officers for the purposes of fisheries legislation.

Look carefully at the powers given to authorised officers. There are two questions to ask of the evidence of the authorised officer.

1. <u>Did he have power to do what he did?</u> Does the Act give power to board as of right or on reasonable suspicion/belief of the commission of an offence? Did the officer have power to take samples, seize exhibits, etc? Did he arrest the vessel or "order it to port"? (This will make a difference if it is later decided not to lay charges).

2. <u>If he didn't, is the ensuing evidence admissible/inadmissible/admissible at the discretion of the judge?</u> In some jurisdictions evidence which is gained unlawfully (e.g. after an illegal search) is inadmissible. There is no discretion in the judge to admit it. In other jurisdictions such evidence might be admissible with the leave of the judge.

A prosecutor should not open or lead evidence which he knows is inadmissible, whose admissibility is subject to the discretion of the judge or to which the defence have given notice it will object.

C. **INTERVIEWS WITH DEFENDANTS**

Recorded interviews with defendants can be a source of very useful evidence in chief and for cross-examination. They can also prove troublesome to an unwary prosecutor and on occasion damage his case far beyond any probative value that might have been gained from them.

The basic rule is they must be voluntary.

The rules concerning interviews are no different in fisheries cases than in any other types of case. They will vary a little from country to country. Set out below is a checklist of the more important rules:-

1. Was the defendant cautioned - and cautioned properly?

2. (a) Was he offered an interpreter?
    (b) Did he understand what was being said?
    (c) Was the interpreter competent and independent?

3. Was he asked if he wished to have a lawyer present? In some jurisdictions there is no need to ask if a lawyer is required. It will generally enhance a prosecutor's submissions that a challenged interview was voluntary if the services of a lawyer have been offered.

4. Was he asked if he wished to have a diplomatic representative present? (If there is not one in the country he should be informed of this).

5. Did the interviewer say who he was and show his authority/identification card?

6. Was he told why he had been arrested/detained?

7. Was he informed of any other matters which are required to be told him by law, (e.g. constitutional rights, if charged or not)?

8. Was he permitted to have another crew member present? This might not be required by law and might in fact hinder the conduct of the interview. However it will help rebut suggestions that a defendant was "on his own in a detention cell with a lot of foreigners and therefore said what he thought they wanted to hear".

9. If there are questions the defendant is obliged by law to answer was he informed of this obligation?

10. Were breaks for rest/refreshment/toilet visits given at reasonable intervals? Were such breaks recorded?

11. (a) Was the interview read back to the defendant at the conclusion?

    (b) Was he asked if it was true and told- he could correct, alter or add anything he wished?

    (c) Were all corrections, alterations or additions initialled?

    (d) Was the interview dated and timed at the beginning and the end?

(e) Was any formal declaration required by law at the beginning and or/end written down and signed by the defendant?

(f) Did the defendant sign at the foot of each page and at the end of the interview?

(g) Did the interviewing officer and any other officials present countersign?

Notes:

(i)   The absence of one or more of the above - listed points will not necessarily render an interview inadmissible. However, where there is a discretion to admit interviews despite defects in the conduct thereof, it will become increasingly difficult to convince a judge that an interview was voluntary the more and the more serious the defects were.

(ii)   A prosecutor should always scrutinize the record of interviews with a defendant carefully and ask himself "Can I obtain evidence to show that anything this defendant has said is untrue?" Such evidence will generally greatly strengthen a prosecutor's case.

(iii)   Remarks made by a defendant other than in a formal interview are generally admissible. Indeed they are often closer to the truth than answers given in a formal interview when the defendant has had time to think out his position. Such remarks might be of great assistance to a prosecution case. However care must be taken in deciding whether or not to lead such evidence, and the following questions should be asked:

(a) were they made at a time when the caution should have already been given?

(b) were they heard by one or more officers (one officer is sufficient to lead such evidence, although he will be uncorroborated on the matter and more open to a "did say/didn't say" argument with the defendant)?

(c) when, if at all, was a note made of the remarks concerned?

(d) was there a likelihood of misunderstanding (e.g. language problems, hearing difficulty, etc.)?

D. **INTERPRETERS**

A prosecutor should compile a list of names, addresses and telephone numbers of people who are willing to act as interpreters, and the languages they speak. They should, if possible, have a knowledge of technical fishing terms. It will help an interpreter employed for a case if he/she has already sat in court as a spectator and familiarised himself/herself with the procedures and court language.

An interpreter should be independent. His/her function is to accurately translate to the defendant everything said in court and everything said by the defendant to the court. Occasionally a witness will require an interpreter to perform the same tasks for him.

The court interpreter should not, if possible, be the same one that was present for the interview; the latter might, in some circumstances, end up as a witness.

Before the trial commences ensure that the interpreter speaks the same language as the defendant. His/her fees should also be agreed in advance - especially if there is no set court scale of fees.

Notes:

(i) It is possible that only one interpreter for a particular language can be found. In those circumstance he/she will have to be the court as well as the interview interpreter. Where there is no interpreter at all the charges should still be proceeded upon subject to the directions of the court. In both circumstances the difficulty should be openly explained to the court.

(ii) A master or other defendant might pretend he knows little or no English when in fact he speaks the language competently. Nearly all radio operators and many masters understand and speak English reasonably well.

(iii) Where a language card has been used to question a defendant (usually upon boarding) then, unless otherwise agreed by the defence, the translation of the questions and correctness of the writing should be proved. This will require someone who knows the language and how it is written, (i.e. someone who will probably qualify as an interpreter).

If answers to language card questions have been written, other than in English, these will also have to be translated.

E. COMMON EXCUSES

Listed below are some of the excuses most often put forward by masters of apprehended vessels. Suggestions on ways to counter them have been added.

1.   "My satellite navigation machine wasn't working"

The authorised officer should have checked this on boarding. Every master of a vessel can navigate by other means (e.g. sextant) or he shouldn't be a master. In most fisheries legislation the fact that the defendant didn't know exactly where he was fishing is irrelevant.

2.   "My radar wasn't working"

The way to counter this is the same as (1), save that radar is only useful for position fixing when a "paint" of land or fixed object of known position can be obtained.

3.   "I thought the Fishing Zone only extended for 12/24/100 miles"

This excuse is becoming a rarity. It is generally known and accepted that 200 miles is the distance states can and have claimed. The master of a fishing vessel will know this. Again the defendant's knowledge of his exact position in any event will probably be irrelevant.

4.   "My engine broke down"

This is a very common excuse, especially if a vessel is stationery when first observed. The authorised officer should as a matter of routine upon boarding ask if the engines are working properly and go and check them. An engineer should board the vessel on arrival in port to also check the engines and look for signs of recent repair, whether or not a breakdown has been alleged. A prosecutor should assess if an engine breakdown is relevant.

5.   "My freezers aren't working"

This excuse is used to explain unfrozen fish in freezers. The same comments apply to this excuse as to number 4. The authorised officer should note if there is a quantity of hard frozen and unfrozen or partly frozen fish in the same freezer.

6.   "I have a licence.....

(a) but its not on board"

Regulations and licence conditions generally require a copy of a licence to be kept aboard the vessel. The authorised officer should have checked by radio, before boarding, any licence numbers or the like painted on the vessel. If this excuse is raised in court the onus is usually, as a matter of law (statutory or evidentiary), upon the defendant to show he had a licence. It would, in any event, be worthwhile having a witness from the fisheries licensing department available to say "I have searched the register and there is no licence/no current licence issued in respect of this vessel".

> Note: It is not often sufficient to rely on reverse onus of proof in a common law court. It has been found, from practical experience, that the prosecution's reliance on the reverse onus of proof has had a tendency to aggravate the court in that it would then appear that the prosecution does not have a case prepared, but is relying on the defendant to convict him/herself.

(b) but I didn't know it had expired/I thought it had been renewed"

Regulations and licence conditions generally require a copy of the current licence to be kept aboard the vessel. Illegal fishing is an offence of strict/absolute liability so even the bona-fide belief that a current licence was held will be no defence if no such licence exists, (see "Law, Strict/Absolute liability").

7.   "I thought you were pirates"

This excuse is given when a master has failed to stop his vessel when required to do so or taken off at full speed. The patrol vessel or other vessel being used for enforcement purposes should have identified itself by radio, be flying appropriate flags, have enforcement officers in uniform and generally comported itself in an official manner.

F. BONDING AND HOLDING

Check in the fisheries legislation if there is a power to bond/a duty to bond a seized vessel and release it, and seize and sell the fish on board. A prosecutor should not be the person to arrange the bond, although he may be called upon to give advice and possibly argue bonding arrangements in court. The United Nations Convention on the Law of the Sea requires that vessels be released upon deposit of a reasonable bond or security.

Unless there is statutory or regulatory provision otherwise, the value of a bond should be the total of:-

(i) a realistic value of the boat, its catch and all its gear, stores and equipment;

(ii) the maximum fine applicable on conviction for the charges laid or to be laid; and

(iii) a reasonable figure in respect of prosecution costs.

It may be, in the absence of statutory provision, that a court will order release upon the lodging of a bond to cover (i). A bond to cover (i), (ii) and (iii) should be required of the owners in the absence of any court direction. The most experienced boat valuer available should be engaged to supply a figure. Boats have a habit of being extraordinarily valuable when first seized yet worth very little when a buy - back price is being negotiated.

Proper arrangements for the lodging of monies, if actually deposited, must be made.

If any part of the boat is required as an exhibit then this should be removed before the vessel is released.

It is not the responsibility of a prosecutor to look after a boat pending a trial. He may however find himself giving advice or appearing in court concerning the matter.

The boat should be kept safely and securely. There are several examples of masters "making a run for it". If essential parts are removed they must be kept safely and provision made for the emergency moving of the boat if adverse conditions arise (e.g. cyclone, storm). The authorities should ensure there is power to remove parts and be wary of the crew buying or manufacturing replacement parts and "making a run for it".

Special care must be taken to ensure that any catch aboard does not spoil. Note any statutory powers there are to sell the catch and hold the proceeds of sale pending the court hearing, and powers to dispose of the catch if it is unsaleable.

Masters or crew members will sometimes sabotage the freezing machinery or other parts of a boat. The holding authorities could then be accused of negligence and incompetence in looking after the boat. Those looking after the boat should be advised to keep records of the checks made and to immediately investigate with an expert any damage or malfunction.

G. ANTECEDENTS

The logs, charts and other documents of the seized vessel should be carefully examined. Much of what is recorded will probably be irrelevant. However these documents might give an indication of the vessel's activities prior to seizure and could well be useful in cross-examination, or for rebutting parts of the defence case.

The full antecedents of the defendants should be obtained in the usual way and, if necessary, checks made with other countries.

The "antecedents" of the vessel concerned should also be obtained, and checks made with other countries.

The information obtained might not be usable in court, even when giving antecedents upon a conviction. However, sometimes such checks can reveal important facts, (e.g. in one Solomon Islands case the@ vessel concerned had only weeks before been arrested for illegal fishing in Papua New Guinea waters, forfeited and sold back).

H. VALUES

A court may well ask the prosecutor questions concerning "value". He should know:-

(i) the value of the vessel, its stores and equipment;

(ii) the value of the vessel's gear;

(iii) the value of the catch (and its weight);

(iv) the value of any bond or security lodged;

(v) the cost of a licence;

(vi) the value, scarcity and growing time of the species taken (e.g.   giant clam);

(vii) the environmental damage of the illegal acts (e.g.   dynamiting fish); and

(viii) the level of profit realisable from the unlawfully   caught fish.

I. THE PRESS

If a foreign fishing vessel is arrested then the local and possibly the international press will be very interested. They will wish to know as much as possible, particularly about any aspects of the incident which will make a "good story". There is generally no harm in keeping the press informed of events. However, the greatest care must be taken if talking to reporters or issuing press statements.

It is far safer for prosecutors and other lawyers who might be involved in a forthcoming trial not to talk to the press. By doing this there cannot be arguments as to whether remarks have been correctly reported or been distorted. It cannot be suggested that a prosecutor has prejudiced a fair trial by some injudicious remark. There is a world of a difference between "the vessel was seized and master arrested for illegal fishing" and "the vessel was seized and master arrested for alleged illegal fishing".

If press releases are to be made they should be written out and carefully considered before issue. Copies must be kept. The release itself should be restricted to the essential facts, contain no statements of opinion and in no way suggest any potential defendant is guilty or wrongful act has been committed.

After a trial is over the same care should be exercised when talking to journalists. Everyone is entitled to a copy of the judgement and transcripts of the evidence. However, opinions should not be expressed to the press as to the correctness of the verdict or adequacy of the sentence.

The best rule is "when in doubt - say nowt".

J. LAW

It need hardly be said that a prosecutor must know thoroughly the law relating to a case which he is prosecuting. This includes not only the entirety of the acts and regulations concerned but also broader background matters, (e.g. rules for hot pursuit, the basis of the 200 mile limit, etc) and subsidiary matters, (e.g. licence conditions).

Special attention should be paid in fisheries cases to:-

1. Jurisdiction

A prosecutor must know the provisions in the legislation which make acts which took place beyond the territory and territorial seas of the country offences and which give courts the jurisdiction to hear those charges. The usual method of extending jurisdiction is to make any offence contrary to the fishing legislation committed within the EEZ or committed anywhere if aboard a domestically registered vessel triable in the country's courts as if that offence had been committed in any place in the country.

The fisheries legislation should also be checked to ensure that proceedings are brought in the right court - Magistrate or High Court, summarily or on indictment/information.

A defence lawyer will always scrutinize legislation in the hope of finding a defect which will result in his client's acquittal or release without the merits of the case being heard.

2. Burden of Proof

Fisheries acts often contain provisions shifting, in specified circumstances, the burden of proof to the defendant e.g. that the defendant did in fact hold a licence. In the absence of statutory provision the standard of proof generally required of a defendant upon whom a burden of proof has been placed is "the balance of probabilities", and not "beyond reasonable doubt".

3. Presumptions

Many statutes create presumptions that will arise once a basis of fact is established (e.g. position entered in official log presumed to be place a vessel was at a particular time unless contrary proved). Use such presumptions, but try not to rely solely upon them. Ensure the presumption evidence contains the necessary basis of fact, and be ready to counter evidence or argument which rebuts a presumption.

Some presumptions will be useable even after conviction, (e.g. on forfeiture all fish found aboard a vessel are presumed to have been caught in the commission of the offence unless the contrary is proved).

4. Strict/Absolute Liability *[noted to raise awareness - this section may not be pertinent in all jurisdictions]*

Many offences created under fisheries acts are ones of strict/absolute liability. Each charge laid must be examined to see if it falls in this category, and argument prepared to support the conclusions reached.

The courts in different jurisdictions approach these offences in two different ways, although all regard them to a greater or lesser degree as negating the necessity for proof of mens rea.

Interpretation 1 - strict and absolute liability are synonymous. If a master is charged-with unlawful fishing all that needs to be proved is that the vessel concerned was fishing within some prohibited area and he is the master of that vessel. It makes no difference whether or not he was mistaken as to his position, area boundaries, etc.

Interpretation 2 - strict and absolute have been given slightly different meanings. An absolute offence is still given the same meaning as it receives under Interpretation 1. However a strict liability offence would admit of a defence if it could be shown by the defendant that he honestly and reasonably believed and had reasonable grounds for believing in a state of facts which, if true, would not be an offence, or he didn't intend to commit an offence and took all reasonable steps to ensure no offence was committed.

The English courts have followed Interpretation I whereas Australian courts have tended to adopt Interpretation 2.

The words "strict" and "absolute" are only labels attached to statutory provisions which are drafted in a particular way. The best approach is to look carefully at the wording of the statute itself in the light of its purpose, the general rules of interpretation and the provisions of any applicable criminal procedure code.

5. International Law

A doctorate in international law is not required to prosecute a fisheries case. However a prosecutor must know the background to, and origins of, the concept of an exclusive economic zone and the quality of, and limitations upon, the exercisable sovereign rights. The meaning of terms such as baseline, territorial sea, internal waters, archipelagic waters should be known. If a bilateral or multilateral treaty is involved then its provisions and interpretations should also be known. The domestic legislation establishing the zone concerned must be available, together with an authentic chart (if such exists) marked with the boundary.

The fisheries acts of some countries forbid the entry of foreign fishing vessels into their EEZ's except for purposes recognised by international law. It would take too long to discuss the purposes for which an entry would be recognised by international law. However on a charge of unauthorized entry it would be for the defendant to put in issue a purpose recognised by international law and for the prosecution to rebut it (unless there is a specific provision shifting the burden to the defendant).

If hot pursuit or cross boundary action is involved, then applicable rules must be known.

6. The Constitution

The vessel involved in the commission of an offence will probably be forfeited upon the conviction of the master or even a crew member. However the owner of the vessel might not be before the court. A prosecutor should check carefully the Constitution of his country to ensure that forfeiture powers are constitutional and are exercised in a way which is consistent with the constitution. The "Deprivation of Property" and "Rights on Trial" sections will probably be the ones relevant to consideration of forfeiture of property.

K. CHARGES

Draft charges with great care - after reading and assessing the evidence and getting to know the law.

There is nothing more embarrassing for a prosecutor than to see a defendant acquitted of Charge A when there was more than enough evidence to convict of Offence B, but it wasn't charged.

The prosecution must prove each element of each charge against each defendant otherwise there will be acquittals.

There are two questions to ask:-

1. Whom do I charge? and
2. With which offences?

1. The Defendant

The master of the vessel concerned should always be charged, unless there are very unusual circumstances.

The vessel owner should also be charged, if this can be done, and the legislation so permits. It does give him a locus standi to argue against forfeiture if there is a conviction.

The fishing master and navigator can, if the legislation so permits, be charged, although this would be unusual. It might in certain circumstances be to the prosecution's advantage to have a fishing master or navigator as a defendant.

It is unusual to charge members of the crew. They will probably not be wealthy, have no control over the boat and its operations and genuinely have no idea where they are. Further, it would appear to be unfair to charge one or two crew members and not the others; if all the crew are charged the trial becomes an unwieldy multi-defendant affair. In some jurisdictions the fisheries legislation does not provide for crew members to be charged.

Crew members are sometimes charged if the case against the master is not very strong. This will happen where legislation so allows, and forfeiture is available or mandatory upon conviction of the master or any member of the crew. The master might be acquitted but the main sanction, forfeiture of the boat, is available upon the conviction of the crew member - the crew-member himself only receiving a small fine.

Note:-

Prosecutors should be wary of one set of circumstances in which the usual illegal fishing provisions ("No foreign fishing vessel shall be used for fishing without a licence" together with "where a vessel is so used the master, etc. shall commit an offence") might cause serious problems.

A master and a crew member could both be charged with illegal fishing and the crew-member has admitted illegal fishing whereas the master has done no more than admit he is the owner of the vessel concerned. There might be some evidence of illegal fishing from the prosecution witnesses but not enough in itself to prove the charge.

At the close of the prosecution the defence could argue the master has no case to answer, and if the trial were to go ahead it would be inviting a conviction of one defendant upon the confession of another. A principle wholly repugnant to the criminal law.

The prosecution could reply it has been proved the vessel has been used for illegal fishing by the general evidence and the crew-member's admission, the master has admitted his position as master and so there is a case to answer.

A court should accept the defence submission - particularly in view of the tenuous nature of the proof and the possibility of a disgruntled crew member wishing to put his master in trouble. If, however, the prosecution submission is rejected a defendant owner or fishing master could argue that a clear and unequivocal admission by a master should not affect him. Further, if the owner were to be acquitted and the master convicted forfeiture of the vessel would evoke a strong protest from the owner.

2. The Charge

The evidence itself will usually suggest the charges that should be laid. It is worth listing all the offences in the fisheries legislation and then checking each in turn against the evidence.

Lesser offences should still be charged even if the evidence clearly reveals serious ones (e.g. a charge of failing to stow gear or unauthorized entry into the EEZ when illegal fishing is the main charge; regulation and licence infringements when a licensed vessel is in a closed area).

Keep clearly in view the potential penalties when deciding upon charges, (e.g. is forfeiture available, will this mean licence cancellation?).

Ensure that each wrongdoing is separately charged and a series of acts do not blur into one charge, (e.g. observed fishing at 6am, 2pm and 6pm on the one day).

Do not be afraid to lay a charge if there are disagreements over the meaning of a section or its applicability. The only way to resolve such disagreements is to argue them before a court and obtain a ruling.

If an authorised officer has been assaulted, obstructed or interfered with in the execution of his duty then charges, whether against the master or crew-members should be laid. This shows that no interference with authorised officers will be accepted, and will also be appreciated by the officers in general. The same approach will apply to such acts as throwing charts overboard, destroying evidence, etc.

It is not the duty of a prosecutor to "charge everything in sight" or "throw the book at a defendant". However, he should charge all offences reasonably disclosed upon the evidence.

L. **BAIL**

Terms of bail are generally set on first appearance in court after arrest and charge. The principal purpose of bail is to ensure the attendance of a defendant at his trial. It would be unusual to request that a foreign national be remanded in custody pending trial. However if there is a real risk that the defendant might abscond with or without his boat and crew then a prosecutor should not flinch from asking for a remand in custody - once he has armed himself with good supporting reasons.

Common terms of bail which are imposed are:

(i) surrender of passport/personal identity documents;
(ii) residence away from boat;
(iii) ban on boarding boat or going to port/jetty area; and
(iv) reporting two or three times a day to a police station.

PRE-TRIAL CHECKLIST

1. Committal Proceedings

Check that:-

    (a)   they have been held, if necessary;

    (b)   the Certificate/Order of commital has been obtained and is in the correct form and signed, stamped;

    (c)   a list of witnesses is with the court;

    (d)   list of exhibits with court; and

    (e)   it is known which exhibits are held by the court, and which by the prosecution.

2. Indictment/Information

    (a)   Check witnesses to be called are all on committal proceedings list. If not, consider notice of additional witness and serve statement on defence. If defendant's statement at committal proceedings can be rebutted then issue notice of additional witness and serve his statement.

    (b)   Check exhibits are all on committal proceedings list. If not, issue notice of additional witness to produce exhibit and serve notice and statement on defence.

    (c)   Review all evidence and check:-

        (i) whether there is evidence to prove every charge against every defendant so charged;
        (ii) if any additional, alternative charges should be laid (and can be done in procedural law); and
        (iii) if any additional defendants should be added (generally possible, but unusual after committal proceedings).

    (d)   Draw up indictment/information.

    (e)   Serve indictment/information on

        (i) court; and
        (ii) all defendants or defence counsel.

3. Witnesses

    Ensure that:-

    (a)   all witnesses on list have been served, and in good time;

    (b)   additional witnesses have been served, and in good time; and any exhibits they produce, obtained and labelled;

    (c)   (i) all necessary experts are witnesses and have been served;
            (ii) statements of experts commence with list of qualifications/experience;
            (iii) their fees are agreed beforehand;

    (d)   there are no problems concerning holidays, illness, unexplained reluctance;

    (e)   (i) transport is provided where appropriate; and
            (ii) any difficulties are known e.g. unreliable inter-island boats.

4. Interpreters

    Ensure that:-

    (a)   they are independent;
    (b)   they are notified and agree to attend court;
    (c)   fees have been fixed (agreed rate or court scale); and
    (d)   they speak the correct language.

5. Exhibits

    Ensure that:

    (a)   all exhibits are in the possession of the court or prosecution;

    (b)   witnesses who are bringing exhibits do so (this should not be a usual practice);

    (c)   documents -

            (i) are originals
            (ii) there are enough copies for judge, prosecution, defence, witnesses and spares for marking.

(d)   perishable exhibits -

(i) are held properly and are ready to be produced if needed (e.g. from freezer);
(ii) proceeds of sale are held, statement of account available.

(e)   for photographs - the police officer/witness has the un-retouched negatives.

6. Certificates

Ensure that:-

(a)   i) all possible evidentiary certificates have been obtained;
ii) they are in correct form and signed;
iii) they have been served on defence, court if required/advisable.

(b)   certificate/documents of appointment of authorised officers involved in case have been checked. (Check applies to right act e.g. not certificate of appointment under Continental Shelf Act when prosecution is under Fisheries Act).

7. Court

Ensure that:-

(a)   the Indictment/Information has been lodged with the Court;
(b)   the case has been listed;
(c)   sufficient number of days have been set aside (i.e. estimate + 50%); and
(d)   the clerk of the Court has been informed there might be many spectators.

8. The Prosecutor

He/she should:-

(a)   know the law thoroughly;
(b)   ave an opening prepared (see "trial" section for details);
(c)   have ensured diplomatic representatives of the defendants have been notified (as appropriate);
(d)   decide if an expert is required to sit with him/her in court (with the court's leave an expert can usually sit with counsel throughout the trial);
(e)   have ready explanatory diagrams, pictures, blackboards etc. and copies, (e.g. for fishing method, navigational system);

(f)   have checked if a certificate or consent of the DPP/AG is needed to prosecute a non-national, it is in the correct form and signed;

(g)   ascertain if the defence are prepared to admit facts e.g. vessel not licensed (N.B. procedure code rules for formal admissions);

(h)   have checked antecedents of defendants, vessel - Criminal Records Office, any other sources of information; and

(i)   know maximum percentage error in normal working of any machines relied upon.

TRIAL

A. OUTSIDE THE COURT

Proofing of witnesses

The practice of proofing a witness is usual in some jurisdictions yet unacceptable in others. Proofing is the reading by a lawyer of a witness statement to the witness before going into court. The word proofing might also cover the testing of the witness on some matters, asking for greater detail, referring to points of importance in the statement and asking him about matters not in the statement but which have become relevant.

The danger is that in proofing a witness ideas, answers and even "facts" which were not in his mind or part of his recollection will be wittingly or unwittingly placed there. And even if no such thing is done or intended the possibility hangs in the air. This is especially so when witnesses are proofed part way through the prosecution case. Defence counsel can ask questions such as "Did you talk to the prosecutor this morning? About the evidence you were going to give? Were you asked about matters not in your statement?" The effect is to make everyone in court wonder whether the prosecutor has been "fixing" his witness or "putting ideas in his head".
However honestly and properly the "proofing" was carried out the witness's standing will be harmed and the prosecutor might find a "bad smell" has attached itself to him.

If the master is foreign then it is probable that a diplomatic representative and members of the press will be present. Their reports will not reflect well upon the country as a whole if the possibility of improper behaviour by the prosecutor exists.

There is no reason why witnesses should not be given their statements to read to refresh their memories before going into court - indeed it is generally a good idea. The object of hearing evidence is not to visit a memory test upon some hapless man or woman. In some jurisdictions the prosecutor may read through the evidence with a witness and ask questions about additional matters or to clarify points.

It is therefore important to know exactly what is or is not permitted by way of proofing a witness.

B. **IN COURT**

1. Interpreters and Experts

The judge or magistrate should as soon as the court is in session be informed of the necessity for an interpreter and the language to be interpreted. With the consent of the court the interpreter should then be sworn in. It is generally best if he sits beside the defendant.

Expert witnesses may, with leave, remain in court throughout the evidence. There are two reasons for this, first to help counsel when examining in chief or cross-examining upon technical matters, and second so they can hear the evidence of witnesses and not need it repeating to them when they come to give their opinions. It is a matter of discretion for the judge or magistrate whether or not to allow an expert to remain in court - the more controversial are the facts to which he is a witness the less likely it is he will be permitted to remain.

The assistance of an expert might be invaluable when it comes to cross examining upon the defence case. He will be able to tell a prosecutor if the defendant is trying to mislead or lie to the court on technical matters.

2. Opening

Every prosecutor has his own particular way of opening a case. It is the moment when the court first hears the details of the allegations and the moment when the prosecutor shapes his case. It is also a time for acquainting the court with the terms of art and technical processes which will be referred to in the evidence.

The interest of the court will wane if an opening commences with a detailed examination of the law or some technical matter. A common form of opening which could well be used in fisheries cases is:-

(a) The Story

Tell chronologically in narrative form what happened. Do not refer to the law or deal with technical matters.

(b) The Technicalities

Explain the meanings of terms of art that will be used, the processes and technical matters involved in the case. This can include not only methods of fishing but details of navigation systems, how they work, slang words used in the industry, types of fish, rates at which fish rot, etc. Use pictures, diagrams and other visual aids.

(c) The Law

Refer to the sections charged, the purpose of the act, evidential burdens, presumptions, whether offences are strict/absolute,

(d) The Defendant

Detail charge by charge, defendant by defendant how the evidence fits each defendant and each charge.

3. Witnesses

(a) Presumptions

If statutory presumptions are to be utilised then ensure that the necessary foundation of fact has been laid by asking witnesses specific questions.

(b) Certificates

Fisheries acts which provide for the proof of facts by certificate generally do not require a witness to produce the certificate in court. It speaks for itself. However if any challenge is made to the certificate or its contents then the maker or someone who can give evidence of the facts in dispute should be called.

Sometimes the veracity or accuracy of the contents of a certificate will not be put in issue until the defence case is presented. If this happens then the court's leave should besought to call or recall a prosecution witness to deal with the matter. The certificate should have been challenged by the defence at the time it was produced to the court.

(c) Experts

After an expert witness is sworn in and has given his name and address he should be asked to give his qualifications and detail his experience in the field in which he professes to be an expert.

Unlike ordinary witnesses, experts can be asked to state opinions upon given sets of facts (e.g. could that fish have been out of the water for more than three days; would this kind of repair require the engine to be shut down/take twenty-four hours; how long had that fish you saw been in that freezer).

(d) Inexperienced Witnesses

(i) Nervousness

Most witnesses called will never or only occasionally have given evidence before. They will be nervous and by reason of that say things they don't mean or know are not correct or omit important matters. Sometimes they might "clam up" and say nothing.

A skilful prosecutor will put his witnesses at ease by asking clear, simple questions on non-contentious matters before dealing with important facts.

A prosecutor will not lightly interrupt a cross-examination. However if he considers a witness is being hectored, intimidated or deliberately confused then a timely intervention to point this out will not be criticised.

Re-examination is a time when confusions and inconsistencies which might have come about in cross-examination can be clarified. A few carefully worded questions can restore the standing of a witness whose reliability, accuracy or veracity has been shaken in cross-examination. Re-examination is not a time for opening new facts or cross-examining one's own witness on evidence he has given which is helpful to the defence.

(ii) Notes

If a witness has made notes contemporaneously with or soon after the events in question then he should bring them to court and use them. The prosecutor should ask the witness early on in his evidence when he made the notes and obtain leave from the court for the witness to use them. Notes are for the purpose of refreshing memory and are not meant to be merely read out.

All fisheries and authorised officers should keep notes whenever they are engaged in the execution of their duties. It must be remembered that most fisheries officers and authorised officers are unlikely to have given evidence more than once or twice before.

(iii) Dress and Behaviour

Courts usually expect witnesses (as far as possible) to be reasonably dressed - and witnesses often feel more comfortable when dressed formally or semi-formally. This does not mean a witness has to go out and buy a suit, but it does exclude holey T-shirts, torn shorts, greasy shirts. Witnesses should also not be chewing gum, betel nut or anything else in court.

It is worthwhile warning inexperienced witnesses of these matters, so they may come prepared.

Fisheries and other authorised officers should wear their uniforms in court (if there is a uniform).

(e) Exhibits

Always maintain a checklist of exhibits and who is producing them. Tick them off as they are produced. Ensure they are all properly labelled.

If a document, such as a chart, is to be marked by a witness in the course of the trial make sure it is a copy. If another witness is to put on the same mark then another and clean copy should be used - to present the first marked copy to the second witness is, in effect, leading him, or her. If it is necessary, an original may, with the court's leave, be marked.

The Court should be requested to view the seized vessel itself if this will help in deciding the case. It is also useful in giving the judge a look at the circumstances in which the boarding and inspection were made.

4. Submissions of "No Case to Answer"

The right of a prosecutor to reply to a submission of no case to answer varies from country to country. In some places there is no limit, whereas in others it is limited to replies upon law only.

Most defence lawyers will not try to misrepresent the law or the evidence. However, if this is done at any time, and particularly upon a submission of no case to answer, it should be firmly and accurately corrected.

Arguments which are wrong in law or fact, in the opinion of the prosecutor, should also be firmly and accurately corrected. The following are examples of erroneous suggestions: that an offence requires mens rea when it is one of strict/absolute liability; that radar is not a notorious scientific instrument; that the readings of a satellite navigation machine are not admissible even if cross-checked against recognised scientific instruments; that the word "now" called out to coincide the activities of two witnesses is hearsay.

5. Cross-Examination

Cross-examination in a fisheries case will be no different from that in other types of cases. It requires the prosecutor to have a thorough knowledge of his own case and evidence and the defence case and evidence. If time permits cross-examination should be prepared point by point. Many books have been written on the art of cross examination.

In the final analysis the effectiveness of a cross-examination is the measure of the skill of the prosecutor. As a broad guide it is better to take a few good points and deal with them rather than embark upon a meandering journey through the defence case.

The prosecutor must have in clear focus the points in dispute between the prosecution and defence. He should then cross-examine thereof and "put" the prosecution case to the defence witnesses. Some examples are set out below.

(a)   The master does not challenge that he was fishing and had no licence but says he was in international waters,

Cross-examination should be directed towards the question of where the vessel was when it fished:-

    (i)   if the master says my satellite navigator was not working, he can be asked "then you don't know exactly where you were?" and "you can't refute the authorised officer's evidence when he says you were 11 miles inside the EEZ?".

    (ii)   if the master says he fixed by sextant at a particular time then calculate the dead reckoned position as adjusted for tides and currents. Cross-examine on this if it puts him inside the EEZ.

    (iii)   if the master says his satellite navigator showed him to be outside the EEZ, put to him that the patrol vessel's satnav after a fix showed him inside the zone/that the boarding party found his satnav not switched off and showed him to be inside the zone.

    (iv)   ask about the rubbed out pencil marks on the chart 11 miles inside the EEZ.

(b)   The master admits he was in the zone but denies he was fishing,

Cross-examination should be directed towards all the indications of recent fishing.

    (i)   why were the scuppers running blood when the boarding party came aboard?

    (ii)   why was the fish that had been lying on the deck for two days still edible raw?

(iii)   how was it that a third of the fish in the port No.2 freezer were not even cold when the rest were rock solid frozen?

iv)   how did the two crew members receive those patterns of superficial cuts on their elbows and knees (clam boats)?

(c)   The master says he genuinely thought the fishing limit was 100 miles,

First, decide if this will be a defence.

Second, cross-examination should be directed to his knowledge as a master:-

    (i)   how much training have you received?

    (ii)   which courses have you been on?

    (iii)   how long have you been a master on fishing vessels?

    (iv)   you have never heard of the 200 mile EEZs in 14 years at sea and four years as a master? that cannot be true?

    (v)   you have heard of the 200 mile EEZ!? then did you check to see if that applied in the case of this country?

d)   The master agrees he took off at speed when approached by the patrol vessel but says he thought it was a pirate ship or unauthorized vessel.

Cross-examination should be directed to the appearance of the patrol vessel and its crew.

    (i)   the patrol vessel's name and number were marked?

    (ii)   it flew the correct flags, didn't it?

    (iii)   the crew were all in uniform?

    (iv)   they identified themselves on the radio?

    (v)   you took off because you wanted to avoid being caught for illegal fishing- isn't that the truth?

Notes:

(i)   If an expert witness is called for the defence then questions can be asked to ascertain if he is truly independent or has some link with the defendant, e.g. Does he work for the company that owns the seized vessel? Is the expert's company partly or wholly financed by the company that owns the seized vessel?

Unless there is very strong evidence to show the expert is not putting forward genuine opinions it should not be suggested. However, if there are strong financial or other links with the defendant or the company that owns the seized vessel then this will allow the court to assess how truly independent the expert is.

(ii)   It may happen that a new field of evidence is opened up during the defence case. If this field of evidence could not reasonably have been forseen and the prosecution can bring or recall witnesses to rebut parts or all of it then a prosecutor should not hesitate to ask leave of the court to bring or recall those witnesses.

6. Closing Speeches

In making a closing speech the prosecutor should refer to the charges to show how every element of every offence has been proved against each defendant. He should briefly deal with those elements which are agreed or over which there is no serious challenge and then deal in detail with the evidence concerning the disputed matters. In doing so points which the defence are likely to make should be forseen and pre-empted. If there are matters of law in contention the prosecutor must state his submissions thereof and how he says the facts fit those interpretations of the law.

If the case against any particular defendant or on any charge is weak then this should be quickly conceded. The prosecutor should cite what supporting evidence there is and then move to the stronger parts of his case.

7. Sentence

The prosecutor must have ready and available:

(a)   the antecedents of all the defendants;
(b)   the antecedents of the vessel;
(c)   the value of the vessel, its stores and equipment;
(d)   the value of the vessel's fishing gear;
(e)   the value of the catch (and its weight);
(f)   the cost of a licence;

(g)   the prevalence of the types of offences concerned;
(h)   the environmental and financial harm caused by the types of wrongful activity of which the defendant has been convicted (including the scarcity and growing time of species illegally taken, e.g. giant clam);
(i)   the maximum fines for each offence;
(j)   the fines imposed in similar cases previously and brief synopses of those cases;
(k)   if possible, the details set out in (j) from other countries;
(l)   for which offences forfeiture is available/discretionary/mandatory and whether for the vessel and/or gear and/or catch;
(m)   the value of any bond held; and
(n)   the likely financial gain the defendant would have made had he not been caught.

Notes:

(i)   If forfeiture is discretionary the court might require a formal request before exercising the power.

(ii)   There might be circumstances where the court has a discretion to forfeit the vessel, gear and catch, but it would be wiser to only confiscate the gear and catch. For example, if the vessel is worth little or is unsaleable and the cost of repatriating the master and crew will fall on the government, then the vessel, less its gear and catch, can be returned so the master and crew can sail home.

(iii) If appeals are contemplated by either the prosecution or defence then application should quickly be made to the court for orders retaining the boat, gear, equipment and catch pending the appeal. Bond or security monies if held should not be paid out, the makers of bonds not released from their obligations and exhibits and perishables preserved.

(iv)   If the master or crew members are fined then the prosecution should ask for restrictions similar to bail conditions to be imposed until the fines have been fully paid.

(v)   Know who is responsible for dealing with a forfeited vessel's gear, equipment and catch. Have him standing by even before the end of the trial. Inform him of any forfeiture as soon as it occurs.

(vi)   Know who is responsible for repatriation of the crew of a forfeited vessel, since they won't be able to pay their own airfares. Inform him of any forfeiture of the vessel as soon as it occurs.

(vii)   Request any necessary orders for the disposal of exhibits.

C. **POST-TRIAL**

1. Keep copies of the charges, synopses of the evidence and judgments.

2. Analyze:-

(a)   your own performance, the mistakes and the good points;

(b)   .the performance of others involved. Discuss with them ways of improving everyone's performance.

4. Ensure the vessel, its gear and catch are returned to the correct person or taken possession of by the correct authority.

5. Check that fines are paid by the time given, and if not institute enforcement proceedings.

6. Ensure that bond or security monies have been returned to the correct person or paid to the correct authority.

7. Arrange for exhibits to be disposed of according to the court's orders.

POSITION FIXING AND NAVIGATION AND LOGS

The position of a vessel at the time of fishing is crucial - was it inside or outside the EEZ, the territorial waters, the closed area?

It is vitally important that the position of the defendant's vessel has been fixed accurately. Even if a vessel is apprehended fishing say fifty miles inside an EEZ the defence may attack the overall reliability of an authorised officer by testing his navigation expertise, in particular the exact location of the vessel when boarded.

Where an apprehension is made near a border line the defence will strongly attack the accuracy and reliability of the prosecution evidence on position. The master might also allege that he didn't know his exact position because his various navigation instruments were not working. The more astute master will assert he was outside the forbidden area.

Fishing must be carried out in the forbidden area for an offence to be committed. For example, if a longliner is apprehended outside or on the border of a forbidden area but its line runs well into it that would constitute illegal fishing.

A. POSITION FIXING AND NAVIGATION

Listed below are the principal instruments and other aids used in the fixing of position and navigation of a vessel.

1. Compass

(a) Magnetic

Every vessel will carry a magnetic compass. It can be used for fixing position by taking bearings off two or more points of land or for running a course.

There are two principal sources of error:-

variation - the earth's magnetic poles do not coincide with the actual poles. The difference between the magnetic pole and the real pole is called the angle of variation. It varies according to position and time. Charts are marked with this angle in degrees east or west, the date on which such angle was calculated and the amount it increases or decreases by each year. Thus by a simple calculation this error can be eliminated.

deviation - the vessel on which a magnetic compass is situated will have its own magnetic field which will affect the magnetic compass. The amount by which this field puts the magnetic compass out is known as the angle of deviation. This angle can be ascertained by a procedure called swinging the ship. It should be done approximately once a year. If the vessel is swung a deviation card recording the reading may be posted near the, chart table. In the case of many fishing vessels this is not so, and thus the master will not accurately know his angle of deviation. It is generally only a few degrees, but could be as much as 20 or 30 degrees.

(b) Gyro

This type of compass relies upon gyroscopes (like very fast spinning tops) and not the magnetic field of the earth. When properly set up dials are adjusted for the latitude and speed of operation. Every vessel will carry a gyro compass.

Every vessel when leaving port should check its gyro compass. Near most ports two large white triangles are permanently set up some distance apart in a prominent place. When the two objects, as viewed from the sea, are directly in line then the gyro compass bearing is noted. The chart for the port in question will show the true bearing of the two objects. Thus any error in the gyro compass can be simply calculated and corrected. This check can be carried out wherever two such triangles are placed - or even by the use of any two suitably placed objects or sharply defined pieces of land which are marked on a chart.

If a gyro compass breaks down out of sight of land then a bearing of the sun called an azimuth can allow it to be set up again with the same accuracy as with land checks. A book of bearings taken to check the gyro might be kept.

The gyro compass can be used for fixing position by taking bearings and will be used for running a course.

2. Radar

In simple terms radar works by sending out an impulse or wave which is reflected when it hits an object. The reflected wave appears as a dot on a screen and shows the relative bearing and distance of the object.

The impulse is emitted by the scanner. This is either a curved, roughly rectangular metal dish or lattice which is placed near the top of the main mast or a one to two meter beam placed in the same area. It rotates and thus "scans" as it goes around and receives reflections. These reflections are relayed to the radar screen, which is found on the bridge or in the chart room. The radar screen is usually circular but may look like a television. A faint line from the centre to the edge rotates as the scanner rotates. Reflections are "painted" as it rotates and gradually fade until the line passes again. An object appearing on the screen is referred to as a "paint" or a "blip". Coastline will appear as a line matching the shape of the coast. The size of the paint will depend upon the distance of the object and its size, although the usual paint is about the size of a pin head. Some small objects will not paint up or only paint when very close. Metallic objects will generally paint up more easily than non-metallic ones.

The range of the radar can be varied so that the distance from the centre of the screen (i.e. where- the ship with the radar is) to the edge is 1,3,6,12,24, or 40 miles etc. depending upon the capability of the radar. "Range rings" can be switched on - these will display on the screen concentric circles; at set distances apart from the centre. A "strobe" can be turned on. This is a screen-displayed dot which goes around. By turning a knob it can be moved outwards or inwards and its distance from the centre read off from a dial. Thus if the distance away of an object must be ascertained the strobe is turned outwards or inwards until it passes over the object; the distance on the dial is then read off.

Radar will show the movements of the radar vessel and any objects painted relatively to each other. Most vessels will use a "grease" pencil to write or mark on the screen objects they wish to track. Thus the course and speed of another vessel might be determined.

Radar can be used to detect vessels, land and objects in the water when visibility is not good. It can be used to monitor the course and speed of another vessel. Heavy seas and heavy rain will interfere with radar by producing a fuzziness on the screen. If another ship is operating its radar in the vicinity it will appear as a "white shadow" passing regularly across a radar screen, or like the "spokes" of a bicycle wheel moving around the screen.

Radar reflectors can be fitted to small boats and buoys. These are simple diamond-shaped metal objects about half a metre across which are fitted to masts. They will show up clearly on a radar screen. A fishing vessel might well use a radar reflector on a buoy attached to a log or at the end of a longline.

3. Satellite Navigation Machine (Satnav)

Most fishing vessels will carry a satnav machine. This machine displays the latitude and longitude of the vessel on a screen.

A series of satellites orbiting the earth each emit signals. The signal of a satellite as it passes over a vessel is received, identified, the Doppler shift measured and the vessel's position displayed. (Doppler shift can crudely be described as the compression or expansion of a wave caused when a wave emitting object is moving towards or away or relative to a wave receiving object. The amount of shift depends upon the relative speed of movement). The receiver is generally a short cylindrical looking object placed on a mast. It is about two-thirds of metre in diameter.

The accuracy of a position given by a satnav machine upon a satellite pass can be within 100 metres. Satellites will pass over a vessel on average about every 90 minutes, although it can be more or less frequent. The largest gap between satellite passes is approximately 3 hours. Sometimes two or three satellites might pass within the space of an hour.

If a vessel is moving then its position must be ascertained by DRing (dead reckoning). This means the vessel's speed and course and any tidal or current set are calculated and plotted on from the known position. Clearly the longer the time lapse from a satnav fix the less accurate will be the calculated position. A vessel's course and speed and tidal and current set can be fed directly or indirectly into the satnav machine and it will, in the light of that data, continuously update and display the vessel's position. The satnav machine itself will over a period calculate any set from current and tide.

Satnav machines are generally very reliable and can withstand temperatures up to 40-45C. Suggestions of satnav malfunction should be examined closely and treated with some scepticism.

The most likely source of error is in the initial setting up of the machine, when a voyage is commenced. However, the error will probably place the vessel in the wrong ocean rather than merely a few miles away.

4. Omega

Omega is commonly referred to as VLF-OMEGA standing for Very Low Frequency. Like a satnav machine Omega receives transmissions and converts them into a position. These transmissions come from nine specially constructed stations world wide and some communication stations. The vessel navigator selects the best stations for the area, using the machine itself to determine signal strength, and the machine will then automatically display the position.

Unlike Satnav, Omega gives a continuous position read-out. Omega is more suited to the higher speed operations of aircraft than surface vessels. In the Pacific Omega has a guaranteed accuracy of better than 2.9 nautical miles, a demonstrated accuracy of 0.5 - 1.5 nautical miles, and when coupled with GPS (Global Positioning System - the new and improved Satnav of the 90's) will have an accuracy within 60 metres.

5. Sextant

This instrument is almost as old as navigation itself. The master of most large DWFN vessels knows how to use a sextant and calculate his position with it, even if he hasn't done so for several years. It is a fairly accurate method of position fixing. The degree of accuracy depends upon the competence of the user. A good operator can expect accuracy within two nautical miles. By use of a split mirror the angle between the horizon and the sun is measured (the moon, stars or planets can be used). The exact moment of doing this in Greenwich Mean Time is noted. This will usually be done at dawn, noon and dusk. By use of tables and formulae the vessel's position can be ascertained.

Note:

A master who alleges he didn't know where he was when he was fishing because his satnav had broken down should be asked what other methods he used to fix his position - particularly by sextant. A denial of knowledge of how to fix by sextant should be treated with extreme scepticism.

6. Direction Finding (DF)

All vessels carry DF equipment. Radio stations or beacons which emit radio signals on set frequencies are maintained in many places. They are located by tuning the DF equipment on board a vessel to the frequency of a beacon. The equipment will then show the actual direction of the beacon from the vessel.

In mid-ocean DF is of no use for position fixing and running a course. Its primary use is for running directly to or from the position of a beacon, for example a beacon attached to a fishing buoy or net or line marker.

7. Echo Sounders/Fish Finders/Sonars

In its simplest terms, an echo sounder is used for measuring the depths of water beneath a vessel. Recordings are made by a sensitive nib on a continuously rolling paper. An echo sounder emits a sound wave which is reflected and received. It is essentially downwards pointing.

Fish finders are basically the same as echo sounders save that they are more sensitive and are capable of detecting and tracing out fish. The fish finder will show when there is a great aggregation of fish beneath the vessel.

8. Navigational Publications

(a)   Charts are, in effect, maps of the sea and coastlines. They show depths of water, shapes of coastlines, reefs, angles of variation, location of navigational lights and what they flash, lines of latitude and longitude and other information.

Courses are drawn and fixes marked on charts. A fishing vessel will mark the position of a log or floating object to which a buoy has been attached. Sometimes its latitude and longitude will just be noted on a piece of paper if illegal fishing is in progress. A longline master or fishing master will generally mark the position of his line when it has been set.

(b) Tide Tables supply the heights of tides and directions of tidal flows, and also the general set of the current in different regions.

(c) Pilot books give a complete range of information for specific ports and restricted areas of waters, (e.g. currents, prevailing winds, hazardous conditions, sunken objects, local navigation rules, port regulations, principal port personnel).

(d) Notices to Mariners (Notams) These are issued at intervals to ensure that mariners are kept informed of changes to charts and other official publications, temporary hazards (e.g. military exercises), new facilities, amended rules and regulations, etc.

9. Inertial Navigation System

This system is only fitted in aircraft. The exact latitude and longitude of the aircraft is entered at a known point (usually the airport of origin of the flight). The system will detect and measure all movements of the aircraft whether along, side to side or up and down. This information is continually plotted on from the aircraft's known starting point. Thus the position of the aircraft during its journey can be read at all times. It is accurate to within a few miles after many hours of flight.

B. LOGS

A vessel will keep "logs" or records of its positions, courses and activities. Listed below are the main logs that will be maintained:-

(a) Vessel Log

This is, in effect, the vessel's diary. Recorded in the log will be a day by day and hour by hour record of the vessel's activities, e.g. dates and times of leaving and entering port, courses run, positions fixed at regular intervals, watch officers, weather, fishing activities, unusual events or observations, general activities.

A vessel log will be used trip after trip until it is full. Occasionally masters will maintain a trip log. This is a log for each trip which is handed in to the vessel owners at the end of the trip and a new one commenced for the next trip.

Note:

When illegal fishing is being undertaken a dummy log is sometimes maintained. This purports to show legal fishing, whilst the real log, which is hidden, shows the places where fishing did in fact take place. Authorised officers should always be aware of this possibility. A good prima facie indication that a log is a dummy is its condition, e.g. trip or fishing activity starts on page 1, fresh un-thumbed condition. The real log will be dog-eared, slightly grubby and probably bear coffee stains.

If a log is suspected to be a dummy, but the real one was not found by the authorised officer, cross examination should be directed towards testing its authenticity: e.g, Why does this log commence 11 days after leaving port? Where is the previous one? Why are all the entries in one coloured ink? Check the alleged distances run. Are they all possible? (One Taiwanese dummy log showed positions which meant the vessel was running at 38 knots for 10 hours. The average fishing boat will run at 8 to 12 knots).

(b)   Fishing Logs are maintained to record all fishing activity: e.g. all positions where fishing was attempted, time, catch by species and weight, weather, method used etc. This information will be later analyzed by the vessel's company and fishing master for future use. Not all vessels maintain a fishing log, the information being recorded in the vessel log.

(c)   Engine Logs record all details of the workings of the engines: e.g. temperatures, pressures, settings for running, breakdown, running maintenance. They are usually kept in the engine room.

An authorised officer should always take possession of the engine log as quickly as possible.

(d)   Freezer Logs are used to record the workings of freezers: e.g. when turned on, temperatures at set times, weight of fish held. An authorised officer should take possession of these logs for investigative purposes.

(e)   Radio Logs record radio messages sent and received (and overheard). They will often be no more than an exercise book ruled into columns. These logs should be checked carefully. Often a vessel will radio its position at regular intervals to a mothership. The notes of these positions should be examined and also compared with positions entered in the vessel log.

C. EVIDENCE FROM INSTRUMENTS OF NAVIGATION AND LOGS

Useful evidence can often be acquired from instruments of navigation, their settings and associated records.

For example:

(a) Compass When a vessel is running, the course to steer might be chalked up for the helmsman on a board near the wheel. Was this the course she was steering when approached? Was this the course she was steering when first picked up on the radar - but subsequently turned to another e.g. to suggest transiting a zone as opposed to running straight into it. The autopilot (device for keeping the vessel on a particular course automatically) might reveal the course a vessel was recently following.

(b) Radar set on 1 mile range when vessel boarded a long way from land. Vessels will usually leave radar on 12 mile range for collision avoidance purposes when steaming. A 1 mile setting suggests a small object is being sought for retrieval e.g. fishing buoy. Any plotting marks on the radar must be examined to ascertain what they refer to.

(c) Satellite Navigation Machines A few vessels carry a printer attached to their satnav to record positions. The print-outs will show where the vessel has been and when. This can be compared with the master's version of events. Periods when the vessel remained in a particular area for a long time can be examined.

The satnav should always be checked as soon as possible upon boarding, and notes made of all readings.

(d) Sextant If a sextant is located then it should be checked for signs of recent use. A stiff sextant with a dust film on the mirrors suggests it has not been used recently.

A stop watch by a sextant or on a chart table could suggest recent use of the sextant. The moment the sun is shot the stopwatch is started so the exact time of the shot can be ascertained after returning to the chart room and looking at the chronometer or accurate clock.

(e) Direction Finding The frequency settings of the DF should be noted upon boarding. Is it tuned to the signal from a radio buoy used in fishing?

(f) Echo Sounder/Fish Finder Does the trace show fish? If so, does the time correspond with indications of fishing from other sources? Does the trace show the vessel has searched for and found a sea mount? Fish are often to be found around sea-mounts (steep rising up of the sea floor) and fishing vessels will seek them out.

Occasionally, and after much work, the trace of the bottom on an echo sounder or fish finder can be matched with the depth markings on a chart to give an exact location of a course the vessel has run.

(g)   Charts may reveal all kinds of useful evidence e.g rubbed-out lines, EEZ boundaries faintly drawn in, position of long-line set marked. Deductions might be made from which chart is out on the table. A master will often put charts as he uses them in the top chart drawer, intending to put them properly away later. The order of charts in the drawer working from the top downwards might give some idea of where the vessel has been.

(h)   Vessel Log This document should always be scrutinised carefully. Is it genuine? Is it consistent with the charts and other logs? Try to read all crossings out. Does it read consistently: e.g. fixes every few hours generally marked then a period of 2 or 3 days when only a few dubious ones are entered?

(i)   Engine Room Log This log might reveal or corroborate by the nature of the entries when a vessel has been fishing e.g. steady speed for a long time, slow speed and then manoeuvring using engines. It might rebut a master's suggestion of engine breakdown.

(j)   Freezer Log A steady temperature of say -18C followed by a quick rise in temperature and gradual return to - I8C is an indicator that fresh fish might have been placed in the freezer.

(k)   Other Indicators Authorised officers should always be alert to and record other navigational indicators: e.g. one side of a vessel is wet from spray and the other relatively dry. Yet on boarding the sea is coming from the dry side. This suggests a recent change of course.

This information has been presented courtesy of the South Pacific Forum Fishery Agency, ICOD and Mr. Robert Coventry.

INTERNATIONAL LAW INFLUENCING FISHERIES MANAGEMENT

The sixtieth party ratified the United Nations Law of the Sea Convention in November 1993. The Convention therefore comes into force in November 1994.

The Convention with its 320 articles and nine annexes is too voluminous to reproduce in this document. This annex is meant to serve as a reference for Fisheries Administrators as to provisions of the Convention of particular relevance to fisheries management and the design and implementation of MCS strategies.

It must be noted that there are other provisions which may have application to fisheries in certain cases consequently, Fisheries Administrators are cautioned not to view this list of provisions as exhaustive.

In essence, the Convention provides principles and rules regarding, but not limited to:

i.	Limits of the territorial sea, contiguous zone, exclusive economic zones and the continental shelf,
ii.	Rights of innocent passage,
iii.	Straits used for international navigation,
iv.	Special circumstances for archipelagic States,
v.	Fishing and passage on the high seas,
vi.	Rights of land-locked States and disadvantageous States.
vii.	Conservation of marine resources, both internal to the EEZs and on the high seas,
viii.	Protection and preservation of the marine environment,
ix.	Exploitation of renewable and non-renewable resources in the EEZs and the sea-bed,
x.	Marine scientific research and technology,
xi.	Pollution,
xii.	Settlements of disputes,
xiii.	Flag State responsibilities,
xiv.	Coastal State responsibilities.

The principal fisheries provisions are on the following pages.

Article 33

Contiguous zone

1.   In a zone contiguous to its territorial sea, described as the contiguous zone, the coastal State may exercise the control necessary to:

     (a)   prevent infringement of its customs, fiscal, immigration or sanitary laws and regulations within its territory or territorial sea;

     (b)   punish the infringement of the above laws and regulations committed within its territory or territorial sea.

2.   The contiguous zone may not extend beyond 24 nautical miles from the baselines from which the breadth of the territorial sea is measured.

Article 55

Specific legal régime of the exclusive economic zone

The exclusive economic zone is an area beyond and adjacent to the territorial sea, subject to the specific legal régime established in this Part, under which the rights and jurisdiction of the coastal State and the rights and freedoms of other States are governed by the relevant provisions of this Convention.

Article 56

Rights, jurisdiction and duties of the coastal State in the exclusive economic zone

1.   In the exclusive economic zone, the coastal State has:

     (a)   sovereign rights for the purposes of exploring and exploiting, conserving and managing the natural resources, whether living or non-living, of the waters superjacent to the sea-bed and of the sea-bed and its subsoil, and with regard to other activities for the economic exploitation and exploration of the zone, such as the production of energy from the water, currents and winds;

     (b)   jurisdiction as provided for in the relevant provisions of this Convention with regard to:

          (i)   the establishment and use of artificial islands, installations and structures;
          (ii)   marine scientific research;
          (iii)   the protection and preservation of the marine environment;

     (c)   other rights and duties provided for in this Convention.

2.   In exercising its rights and performing its duties under this Convention in the exclusive economic zone, the coastal State shall have due regard to the rights and duties of other States and shall act in a manner compatible with the provisions of this Convention.

3.   The rights set out in this article with respect to the sea-bed and subsoil shall be exercised in accordance with Part VI.

Article 57

Breadth of the exclusive economic zone

The exclusive economic zone shall not extend beyond 200 nautical miles from the baselines from which the breadth of the territorial sea is measured.

Article 58

Rights and duties of other States in the exclusive economic zone

1.   In the exclusive economic zone, all States, whether coastal or land-locked, enjoy, subject to the relevant provisions of this Convention, the freedoms referred to in article 87 of navigation and overflight and of the laying of submarine cables and pipelines, and other internationally lawful uses of the sea related to these freedoms, such as those associated with the operation of ships, aircraft and submarine cables and pipelines, and compatible with the other provisions of this Convention.

2.   Articles 88 to 115 and other pertinent rules of international law apply to the exclusive economic zone in so far as they are not incompatible with this Part.

3.   In exercising their rights and performing their duties under this Convention in the exclusive economic zone, States shall have due regard to the rights and duties of the coastal State and shall comply with the laws and regulations adopted by the coastal State in accordance with the provisions of this Convention and other rules of international law in so far as they are not incompatible with this Part.

Article 61

Conservation of the living resources

1.   The coastal State shall determine the allowable catch of the living resources in its exclusive economic zone.

2.   The coastal State, taking into account the best scientific evidence available to it, shall ensure through proper conservation and management measures that the maintenance of the living resources in the exclusive economic zone is not endangered by over-exploitation. As appropriate, the coastal State and competent international organizations, whether subregional, regional or global, shall co-operate to this end.

3.   Such measures shall also be designed to maintain or restore populations of harvested species at levels which can produce the maximum sustainable yield, as qualified by relevant environmental and economic factors, including the economic needs of coastal fishing communities and the special requirements of developing States, and taking into account fishing patterns, the interdependence of stocks and any generally recommended international minimum standards, whether subregional, regional or global.

4.   In taking such measures the coastal State shall take into consideration the effects on species associated with or dependent upon harvested species with a view to maintaining or restoring populations of such associated or dependent species above levels at which their reproduction may become seriously threatened.

5.   Available scientific information, catch and fishing effort statistics, and other data relevant to the conservation of fish stocks shall be contributed and exchanged on a regular basis through competent international organizations, whether subregional, regional or global, where appropriate and with participation by all States concerned, including States whose nationals are allowed to fish in the exclusive economic zone.

Article 62

Utilization of the living resources

1.   The coastal State shall promote the objective of optimum utilization of the living resources in the exclusive economic zone without prejudice to article 61.

2.   The coastal State shall determine its capacity to harvest the living resources of the exclusive economic zone. Where the coastal State does not have the capacity to harvest the entire allowable catch, it shall, through agreements and other arrangements and pursuant to the terms, conditions, laws and regulations referred to in paragraph 4, give other States access to the surplus of the allowable catch, having particular regard to the provisions of articles 69 and 70, especially in relation to the developing States mentioned therein.

3.   In giving access to other States to its exclusive economic zone under this article, the coastal State shall take into account all relevant factors, including, inter alia, the significance of the living resources of the area to the economy of the coastal State concerned and its other national interests, the provisions of articles 69 and 70, the requirements of developing States in the subregion or region in harvesting part of the surplus and the need to minimize economic dislocation in States whose nationals have habitually fished in the zone or which have made substantial efforts in research and identification of stocks.

4.   Nationals of other States fishing in the exclusive economic zone shall comply with the conservation measures and with the other terms and conditions established in the laws and regulations of the coastal State. These laws and regulations shall be consistent with this Convention and may relate, inter alia, to the following:

   (a)   licensing of fishermen, fishing vessels and equipment, including payment of fees and other forms of remuneration, which, in the case of developing coastal States, may consist of adequate compensation in the field of financing, equipment and technology relating to the fishing industry;

   (b)   determining the species which may be caught, and fixing quotas of catch, whether in relation to particular stocks or groups of stocks or catch per vessel over a period of time or to the catch by nationals of any State during a specified period;

   (c)   regulating seasons and areas of fishing, the types, sizes and amount of gear, and the types, sizes and number of fishing vessels that may be used;

   (d)   fixing the age and size of fish and other species that may be caught;

(e)   specifying information required of fishing vessels, including catch and effort statistics and vessel position reports;

(f)   requiring, under the authorization and control of the coastal State, the conduct of specified fisheries research programmes and regulating the conduct of such research, including the sampling of catches, disposition of samples and reporting of associated scientific data;

(g)   the placing of observers or trainees on board such vessels by the coastal State;

(h)   the landing of all or any part of the catch by such vessels in the ports of the coastal State;

(i)   terms and conditions relating to joint ventures or other co-operative arrangements;

(j)   requirements for the training of personnel and the transfer of fisheries technology, including enhancement of the coastal State's capability of undertaking fisheries research;

(k)   enforcement procedures.

5.   Coastal States shall give due notice of conservation and management laws and regulations.

Article 63

Stocks occurring within the exclusive economic zones of two or more coastal States or both within the exclusive economic zone and in an area beyond and adjacent to it

1.   Where the same stock or stocks of associated occur within the exclusive economic zone of two or more coastal States, these States shall seek, either directly or through appropriate subregional or regional organizations, to agree upon the measures necessary to co-ordinate and ensure the conservation and development of such stocks without prejudice to the other provisions of this Part.

2.   Where the same stock or stocks of associated species occur both within the exclusive economic zone and in an area beyond and adjacent to the zone, the coastal State and the States fishing for such stocks in the adjacent area shall seek, either directly or through appropriate subregional or regional organizations, to agree upon the measures necessary for the conservation of these stocks in the adjacent area.

Article 64

Highly migratory species

1.   The coastal State and other States whose nationals fish in the region for the highly migratory species listed in Annex I shall co-operate directly or through appropriate international organizations with a view to ensuring conservation and promoting the objective of optimum utilization of such species throughout the region, both within and beyond the exclusive economic zone. In regions for which no appropriate international organization exists, the coastal State and other States whose nationals harvest these species in the region shall co-operate to establish such an organization and participate in its work.

2.   The provisions of paragraph 1 apply in addition to the other provisions of this Part.

Article 65

Marine mammals

Nothing in this Part restricts the right of a coastal State or the competence of an international organization, as appropriate, to prohibit, limit or regulate the exploitation of marine mammals more strictly than provided for in this Part. States shall co-operate with a view to the conservation of marine mammals and in the case of cetaceans shall in particular work through the appropriate international organizations for their conservation, management and study.

Article 66

Anadromous stocks

1.   States in whose rivers anadromous stocks originate shall have the primary interest in and responsibility for such stocks.

2.   The State of origin of anadromous stocks shall ensure their conservation by the establishment of appropriate regulatory measures for fishing in all waters landward of the outer limits of its exclusive economic zone and for fishing provided for in paragraph 3(b). The State of origin may, after consultations with other States referred to in paragraphs 3 and 4 fishing these stocks, establish total allowable catches for stocks originating in its rivers.

3.   (a)   Fisheries for anadromous stocks shall be conducted only in waters landward of the outer limits of exclusive economic zones, except in cases where this provision would result in economic dislocation for a State other than the State of origin. With respect to such fishing beyond the outer limits of the exclusive economic zone, States concerned shall maintain consultations with a view to achieving agreement on terms and conditions of such fishing giving due regard to the conservation requirements and the needs of the State of origin in respect of these stocks.

        (b)   The State of origin shall co-operate in minimizing economic dislocation in such other States fishing these stocks, taking into account the normal catch and the mode of operations of such States, and all areas in which such fishing occurred.

        (c)   States referred to in subparagraph (b), participating by agreement with the State of origin in measures to renew anadromous stocks, particularly by expenditures for that purpose, shall be given special consideration by the State of origin in the harvesting of stocks originating in its rivers.

        (d)   Enforcement of regulations regarding anadromous stocks beyond the exclusive economic zone shall be by agreement between the State of origin and the other States concerned.

4.   In cases where anadromous stocks migrate into or through the waters landward of the outer limits of the exclusive economic zone of a State other than the State of origin, such State shall co-operate with the State of origin with regard to the conservation and management of such stocks.

5.   The State of origin of anadromous stocks and other States fishing these stocks shall make arrangements for the implementation of the provisions of this article, where appropriate, through regional organizations.

Article 67

Catadromous species

1.   A coastal State in whose waters catadromous species spend the greater part of their life cycle shall have the responsibility for the management of these species and shall ensure the ingress and egress of migrating fish.

2.   Harvesting of catadromous species shall be conducted only in waters landward of the outer limits of exclusive economic zones. When conducted in the exclusive economic zones, harvesting shall be subject to this article and the other provisions of this Convention concerning fishing in these zones.

3.   In cases where catadromous fish migrate through the exclusive economic zone of another State, whether as juvenile or maturing fish, the management, including harvesting, of such fish shall be regulated by agreement between the State mentioned in paragraph 1 and the other State concerned. Such agreement shall ensure the rational management of the species and take into account the responsibilities of the State mentioned in paragraph 1 for the maintenance of these species.

Article 73

Enforcement of laws and regulations on the coastal State

1.   The coastal State may, in the exercise of its sovereign rights to explore, exploit, conserve and manage the living resources in the exclusive economic zone, take such measures, including boarding, inspection, arrest and judicial proceedings, as may be necessary to ensure compliance with the laws and regulations adopted by it in conformity with this Convention.

2.   Arrested vessels and their crews shall be promptly released upon the posting of reasonable bond or other security.

3.   Coastal State penalties for violations of fisheries laws and regulations in the exclusive economic zone may not include imprisonment, in the absence of agreements to the contrary by the States concerned, or any other form of corporal punishment.

4.   In cases of arrest or detention of foreign vessels the coastal State shall promptly notify the flag State, through appropriate channels, of the action taken and of any penalties subsequently imposed.

Article 86

Application of the provisions of this Part

The provisions of this Part apply to all parts of the sea that are not included in the exclusive economic zone, in the territorial sea or in the internal waters of a State, or in the archipelagic waters of an archipelagic State. This article does not entail any abridgement of the freedoms enjoyed by all States in the exclusive economic zone in accordance with article 58.

Article 87

Freedom of the high seas

1.   The high seas are open to all States, whether coastal or land-locked. Freedom of the high seas is exercised under the conditions laid down by this Convention and by other rules of international law. It comprises, <u>inter alia</u>, both for coastal and land-locked States:

(a)   freedom of navigation;
(b)   freedom of overflight;
(c)   freedom to lay submarine cables and pipelines, subject to Part VI;
(d)   freedom to construct artificial islands and other installations permitted under international law, subject to Part VI;
(e)   freedom of fishing, subject to the conditions laid down in section 2;
(f)   freedom of scientific research, subject to Parts VI and XIII.

2.   These freedoms shall be exercised by all States with due regard for the interests of other States in their exercise of the freedom of the high seas, and also with due regard for the rights under this Convention with respect to activities in the Area.

Article 95

Immunity of warships on the high seas

Warships on the high seas have complete immunity from the jurisdiction of any State other than the flag State.

Article 96

Immunity of ships used only on government non-commercial service

Ships owned or operated by a State and used only on government non-commercial service shall, on the high seas, have complete immunity from the jurisdiction of any other State other than the flag State.

Article 101

Definition of piracy

Piracy consists of any of the following acts:

(a)   any illegal acts of violence or detention, or any act of depredation, committed for private ends by the crew or the passengers of a private ship or a private aircraft, and directed:

(i)   on the high seas, against another ship or aircraft, or against persons or property on board such ship or aircraft;
(ii)   against a ship, aircraft, persons or property in a place outside the jurisdiction of any State;

(b)   any act of voluntary participation in the operation of a ship or of an aircraft with the knowledge of facts making it a pirate-ship or aircraft;

(c)   any act of inciting or of intentionally facilitating an act described in subparagraph (a) or (b).

Article 102

Piracy by a warship, government ship or government aircraft whose crew has mutinied

The acts of piracy, as defined in article 101, committed by a warship, government ship or government aircraft whose crew has mutinied and taken control of the ship or aircraft are assimilated to acts committed by a private ship or aircraft.

Article 103

Definition of a pirate ship or aircraft

A ship or aircraft is considered a pirate ship or aircraft if it is intended by the persons in dominant control to be used for the purpose of committing one of the acts referred to in article 101. The same applies if the ship or aircraft has been used to commit any such act, so long as it remains under the control of the persons guilty of that act.

Article 104

Retention or loss of the nationality of a pirate ship or aircraft

A ship or aircraft may retain its nationality although it has become a pirate ship or aircraft. The retention or loss of nationality is determined by the law of the State from which such nationality was derived.

Article 105

Seizure of a pirate ship or aircraft

On the high seas, or in any other place outside the jurisdiction of any State, every State may seize a pirate ship or aircraft, or a ship or aircraft taken by piracy and under the control of pirates, and arrest the persons and seize the property on board. The courts of the State which carried out the seizure may decide upon the penalties to be imposed, and may also determine the action to be taken with regard to the ships, aircraft or property, subject to the rights of third parties acting in good faith.

Article 106

Liability for seizure without adequate grounds

Where the seizure of a ship or aircraft on suspicion of piracy has been effected without adequate grounds, the State making the seizure shall be liable to the State the nationality of which is possessed by the ship or aircraft for any loss or damage caused by the seizure.

Article 107

Ships and aircraft which are entitled to seize on account of piracy

A seizure on account of piracy may be carried out only by warships or military aircraft, or other ships or aircraft clearly marked and identifiable as being on government service and authorized to that effect.

Article 110

Right of visit

1.   Except where acts of interference derive from powers conferred by treaty, a warship which encounters on the high seas a foreign ship, other than a ship entitled to complete immunity in accordance with articles 95 and 96, is not justified in boarding it unless there is reasonable ground for suspecting that:

    (a)   the ship is engaged in piracy;
    (b)   the ship is engaged in the slave trade;
    (c)   the ship is engaged in unauthorized broadcasting and the flag State of the warship has jurisdiction under article 109;
    (d)   the ship is without nationality; or
    (e)   though flying a foreign flag or refusing to show its flag, the ship is, in reality, of the same nationality as the warship.

2.   In the cases provided for in paragraph 1, the warship may proceed to verify the ship's right to fly its flag. To this end, it may send a boat under the command of an officer to the suspected ship. If suspicion remains after the documents have been checked, it may proceed to a further examination on board the ship, which must be carried out with all possible consideration.

3.   If the suspicions prove to be unfounded, and provided that the ship boarded has not committed any act justifying them, it shall be compensated for any loss or damage that may have been sustained.

4.   These provisions apply mutatis mutandis to military aircraft.

5.   These provisions also apply to any other duly authorized ships or aircraft clearly marked and identifiable as being on government service.

Article 111

Right of hot pursuit

1.   The hot pursuit of a foreign ship may be undertaken when competent authorities of the coastal State have good reason to believe that the ship has violated the laws and regulations of that State. Such pursuit must be commenced when the foreign ship or one of its boats is within the internal waters, the archipelagic waters, the territorial sea or the contiguous zone of the pursuing State, and may only be continued outside the territorial sea or contiguous zone if the pursuit has not been interrupted. It not is necessary that, at the time when the foreign ship within the territorial sea or contiguous zone receives the order to stop, the ship giving the order should likewise be within the territorial sea or the contiguous zone. If the foreign ship is within a contiguous zone as defined in article 33, the pursuit may only be undertaken if there has been a violation of the rights for the protection of which the zone was established.

2.   The right of hot pursuit shall apply _mutatis mutandis_ to violations in the exclusive economic zone or on the continental shelf, including safety zones around continental shelf installations, of the laws and regulations of the coastal State applicable in accordance with this Convention to the exclusive economic zone or the continental shelf, including such safety zones.

3.   The right of hot pursuit ceases as soon as the ship pursued enters the territorial sea of its own State or of a third State.

4.   Hot pursuit is not deemed to have begun unless the pursuing ship has satisfied itself by such practicable means as may be available that the ship pursued or one of its boats or other craft working as a team and using the ship pursued as a mother ship is within the limits of the territorial sea, or, as the case may be, within the contiguous zone or the exclusive economic zone or above the continental shelf. The pursuit may only be commenced after a visual or auditory signal to atop has been given at a distance which enables it to be seen or heard by the foreign ship.

5.   The right of hot pursuit may be exercised only by warships or military aircraft, or other ships or aircraft clearly marked and identifiable as being on government service and authorized to that effect.

6.   Where hot pursuit is effected by an aircraft:

    (a)   the provisions of paragraph 1 to 4 shall apply _mutatis mutandis_;

    (b)   the aircraft giving the order to stop must itself actively pursue the ship until a ship or another aircraft of the coastal State, summoned by the aircraft, arrives to take over the pursuit, unless the aircraft is itself able to arrest the ship. It does not suffice to justify an arrest outside the territorial sea that the ship was merely sighted by the aircraft as an offender or suspected offender, if it was not both ordered to stop and pursued by the aircraft itself or other aircraft or ships which continue the pursuit without interruption.

7.   The release of a ship arrested within the jurisdiction of a State and escorted to a port of that State for the purposes of an inquiry before the competent authorities may not be claimed solely on the ground that the ship, in the course of its voyage, was escorted across a portion of the exclusive economic zone or the high seas, if the circumstances rendered this necessary.

8.   Where a ship has been stopped or arrested outside the territorial sea in circumstances which do not justify the exercise of the right of hot pursuit, it shall be compensated for any loss or damage that may have been thereby sustained.

Article 116

Right to fish on the high seas

All States have the right for their nationals to engage in fishing on the high seas subject to:

(a)   their treaty obligations;

(b)   the rights and duties as well as the interests of coastal States provided for, inter alia, in article 63, paragraph 2, and articles 64 to 67; and

(c)   the provisions of this section.

Article 117

Duty of States to adopt with respect to their nationals measures for the conservation of the living resources of the high seas

All States have the duty to take, or to co-operate with other States in taking, such measures for their respective nationals as may be necessary for the conservation of the living resources of the high seas.

Article 118

Co-operation of States in the conservation and management of living resources

States shall co-operate with each other in the conservation and management of living resources in the areas of the high seas. States whose nationals exploit identical living resources, or different living resources in the same area, shall enter into negotiations with a view to taking the measures necessary for the conservation of the living resources concerned. They shall, as appropriate, co-operate to establish subregional or regional fisheries organizations to this end.

Article 119

Conservation of the living resources of the high seas

1.   In determining the allowable catch and establishing other conservation measures for the living resources in the high seas, States shall:

(a)   take measures which are designed, on the best scientific evidence available to the States concerned, to maintain or restore populations of harvested species at levels which can produce the maximum sustainable yield, as qualified by relevant environmental and economic factors, including the special requirements of developing States, and taking into account fishing patterns, the interdependence of stocks and any generally recommended international minimum standards, whether subregional, regional or global;

(b)   take into consideration the effects on species associated with or dependent upon harvested species with a view to maintaining or restoring populations of such associated or dependent species above levels at which their reproduction may become seriously threatened.

214

2.   Available scientific information, catch and fishing effort statistics, and other data relevant to the conservation of fish stocks shall be contributed and exchanged on a regular basis through competent international organizations, whether subregional, regional or global, where appropriate and with participation by all States concerned.

3.   States concerned shall ensure that conservation measures and their implementation do not discriminate in form or in fact against the fishermen of any State.

Article 120

Marine mammals

Article 65 also applies to the conservation and management of marine mammals in the high seas.

Article 192

States have the obligation to protect and preserve the marine environment.

Article 194

Measures to prevent, reduce and control pollution of the marine environment

1.   States shall take, individually or jointly as appropriate, all measures consistent with this Convention that are necessary to prevent, reduce and control pollution of the marine environment from any source, using for this purpose the best practical means at their disposal and in accordance with their capabilities, and they shall endeavour to harmonize their policies in this connection.

Article 197

Co-operation on a global or regional basis

States shall co-operate on a global basis and, as appropriate, on a regional basis, directly or through competent international organizations, in formulating and elaborating international rules, standards and recommended practices and procedures consistent with this Convention, for the protection and preservation of the marine environment, taking into account characteristic regional features.

In addition, a list of Migratroy Species:

1. Albacore tuna: Thunnus alalunga.

2. Bluefin tuna: Thunnus thynnus.

3. Bigeye tuna: Thunnus obesus.

4. Skipjack tuna: Katsuowonus pelamis.

5. Yellowfin tuna: Thunnus albacares.

6. Blackfin tuna: Thunnus atlanticus.

7. Little tuna: Euthynnus alletteratus; Euthynnus affinis.

8. Southern bluefin tuna: Thunnus maccoyii.

9. Frigate mackerel: Auxis thazard; Auxis rochei.

10. Pomfrets: Family Bramidae.

11. Marlins: Tetrapturus angustirostris; Tetrapturus belone; Tetrapturus pfluegeri; Tetrapturus albidus; Tetrapturus audax; Tetrapturus georgei; Makaira mazara; Makaira indica; Makaira nigricans.

12. Sail-fishes: Istiophorus platypterus; Istiophorus albicans.

13. Swordfish: Xiphias gladius.

14. Sauries: Scomberesox saurus; Cololabis saira; Cololabis adocetus; Scomberesox saurus scombroides.

15. Dolphin: Coryphaena hippurus; Coryphaena equiselis.

16. Ocean sharks: Hexanchus griseus; Cetorhinus maximus; Family Alopiidae; Rhincodon typus; Family Carcharhinidae; Family Sphyrnidae; Family Isurida.

17. Cetaceans: Family Physeteridae; Family Balaenopteridae; Family Balaenidae; Family Eschrichtiidae; Family Monodontidae; Family Ziphiidae; Family Delphinidae.

TABLE I

EFFECTIVENESS OF DIFFERENT TYPES OF VESSEL MONITORING

Type of MCS	Description of Monitoring	Effectiveness of Monitoring of				Amount of Time Observed	Effectiveness of Detection of Unlicensed Vessels	Coverage at Sea	Power of Arrest
		Position	Fishing Gear	Catches/ Quotas	Days at Sea				
By Vessel	Identification by sight and boarding for Inspection	High	High	Medium	High*	Low	High	300 sq mls per hr	Yes
By Air	Limited to rapidly changing identification	High	Low	None	High*	Low	High	3000 sq mls per hr	No
Shore Based	Inspection of catch and fishing gear. Coastal Surveillance	None	High	High	High	Medium	Low	None	Yes
Observers on Vessels	Continual observation of activities	High	High	High	High	High	Medium	High	No
Vessel Monitoring System	Periodic Monitoring of Vessels Position	High	None	None	High	High	None	Complete for Vessels Fitted	No

*   but depends on frequency of flights or days at sea of monitoring vessel

TABLE II

ADVANTAGES AND DISADVANTAGES OF THE VARIOUS MCS SYSTEMS

Type of MCS	ADVANTAGES	DISADVANTAGES
By Vessel	Provides at sea verification that fishing gear and catch is legal. Most important to control Transhipment and arrest particularly of foreign vessels	Very costly
By Air	Can provide the best coverage for identification of illegal incursion of unlicensed vessels and also of observation of boxes	Very costly. No ability to arrest. No ability to inspect the catch or fishing gear.
Shore based	Lowest running costs and low capital costs. Can monitor landed catch and quotas. Only power of arrest in port.	No possibility of monitoring foreign vessels that do not call at port. No possibility of monitoring transhipment.
Observers	Can observe all operations.	High cost. Only viable on larger vessels.
Vessel Monitoring System	Provides almost real time monitoring of position for fitted vessels and can reduce interception times for enforcement craft. Relatively low capital cost and running costs borne by fishing vessel.	No coverage for vessels not fitted with the system. Involves cost of installation on the fishing vessel. No detection of unlicensed vessels.

NOTE: An MCS System, to be effective, utilizes most of these components if they are available. The emphasis depends on the fisheries and fishing operations of each country, the political support for conservation and the funding available, to note a few factors. The emphasis on one or other component may therefore change according to the situation of the day. This comparison is not to suggest that MCS activities could be done exclusively with one component, the choice open to the Fishery Administrator is the appropriate mix to address their country's fisheries priorities.